GUIDE TO ABSTRACT ALGEBRA

Consultant Editor: David A. Towers, Senior Lecturer in Mathematics, University of Lancaster

Titles available:
  Abstract Algebra
  Linear Algebra
  Analysis

Further titles are in preparation

Guide to Abstract Algebra

Carol Whitehead

Senior Lecturer in Mathematics
Goldsmiths' College
University of London

MACMILLAN
EDUCATION

First published 1988

Published by
MACMILLAN EDUCATION LTD
Houndmills, Basingstoke, Hampshire RG21 2XS
and London
Companies and representatives
throughout the world

Printed in Hong Kong

British Library Cataloguing in Publication Data
Whitehead, Carol
Guide to abstract algebra.—(Macmillan
mathematical guides)
1. Algebra, Abstract
I. Title
512'.02   QA162
ISBN 0–333–42657–6

To my Parents

CONTENTS

FOREWORD TO THE SERIES

Wide concern has been expressed in tertiary education about the difficulties experienced by students during their first year of an undergraduate course containing a substantial component of mathematics. These difficulties have a number of underlying causes, including the change of emphasis from an algorithmic approach at school to a more rigorous and abstract approach in undergraduate studies, the greater expectation of independent study, and the increased pace at which material is presented. The books in this series are intended to be sensitive to these problems.

The books are carefully selected, short, introductory texts on key areas of the first-year syllabus, which are complementary and largely self-contained. Throughout, the pace of development is gentle, sympathetic and carefully motivated. Clear and detailed explanations are provided, and important concepts and results are stressed.

As mathematics is a practical subject which is best learned by doing it, rather than by either watching or reading about someone else doing it, a particular effort has been made to include a plentiful supply of worked examples together with appropriate exercises ranging in difficulty from the straightforward to the challenging.

When fellwalking, the most breathtaking views require some expenditure of effort in order to gain access to them; nevertheless, the peak is more likely to be reached if a gentle and interesting route is chosen. The mathematical peaks attainable in these books are every bit as exhilarating, the paths are as gentle as we could find, and the interest and expectation are maintained throughout to prevent the spirits from flagging on the journey.

David Towers
Consultant Editor

To encourage the reader who finds abstract ideas difficult to grasp, I have tried to illustrate every new concept by concrete examples to show how the idea works or how the new symbol or term is used. A conscious effort has been made to introduce these gradually so that each new idea is well illustrated and practised before the next is introduced. At the end of every section there is a set of exercises for which fairly full solutions are provided. Although some of these problems extend the material of the text, their chief aim is to develop the reader's mastery of the ideas and techniques just presented and to build his or her confidence in solving problems and expressing their solution in a clear and mathematical way. To this end, many of the exercises are of a quite straightforward nature. There is an additional set of problems at the end of most sections for which the solutions are not given.

This book was written as a bridging book between school and university mathematics and so should be very suitable for a sixth-form readership.

Many people have helped me with the preparation of the manuscript. Dr David Towers read the finished manuscript with meticuluous attention and pointed out a number of errors and suggested many improvements. His help has been invaluable and I cannot thank him sufficiently. I must also express my gratitude to Dr Bill Jackson and Dr Anthea Grubb for class testing parts of the material and making a number of helpful suggestions; to Mrs Elsie Lockton and Mrs Barbara Brignall for typing the manuscript and to many of my students, past and present, for their suggestions and encouragement.

Goldsmiths' College, University of London, 1988   C. W.

GLOSSARY OF SYMBOLS

Symbol
or
notation

Meaning

Set notation

$\{x: p(x)\}$ The set of all elements x for which the statement $p(x)$ is true.

$x \in A$ The element x **belongs** to the set A.

$A \subseteq B$ The set A **is contained in** the set B.

$A \subset B$ The set A **is a proper subset of** the set B.

$A \supseteq B$ The set A **contains** the set B **as a subset**.

$A \supset B$ The set A **contains** the set B **as a proper subset**.

$|A|$ The **number of** elements in the finite set A.

$\varnothing$ The empty set.

$\mathscr{E}$ The universal set.

A' The **set complement** of the set A.

$\mathscr{P}(A)$ The **power set** (set of all subsets) of A.

$A \cup B$ The **union of** sets A and B.

$A \cap B$ The **intersection of** sets A and B.

Symbol or notation	Meaning

Set notation (cont.)

$A - B$	The **set difference** of sets A and B.
$A \triangle B$	The **symmetric difference** of sets A and B.
$A \times B$	The **cartesian product** of sets A and B.

Other symbols

$\mathbb{Z}$	Set of integers.
$\mathbb{Q}$	Set of rational numbers.
$\mathbb{R}$	Set of real numbers.
$\mathbb{C}$	Set of complex numbers
S^+	Set of strictly positive members of a set S.
S^*	Set of non-zero members of a set S.
$\mathbb{Z}_n$	Set of congruence classes modulo n.
S_n	Symmetric group on n symbols.
A_n	Alternating group on n symbols.
D_n	Dihedral group of order $2n$.
C_n	Cyclic group of order n.
V	Klein 4-group.
ι	Identity permutation
i	A complex square root of -1.
$\Rightarrow$	Implies that.
$\Leftarrow$	Is implied by.

Symbol or notation	Meaning
$\Leftrightarrow$	If and only if.
$\forall$	For every.
$\oplus$	Addition of congruence classes.
$\otimes$	Multiplication of congruence classes.
$a\ \boldsymbol{R}\ b$	The element a **is related to** the element b by the relation $\boldsymbol{R}$.
$a \sim b$	The element a **is related to** the element b by an equivalence relation $\sim$.
$a \equiv b \pmod{m}$	The integer a **is congruent to** the integer b modulo m.
$G \cong H$	The group G **is isomorphic to** the group H.
$\bar{a}$	The equivalence (or congruence) class of a.
$a\|b$	The integer a **is a factor of** the integer b.
$o(a)$	The order of an element a of a group.
$\langle a \rangle$	The cyclic group generated by the element a.
$\langle a_1, a_2, \ldots, a_m \rangle$	The group generated by the elements $a_1, a_2, \ldots, a_m$.
$(a_1, a_2, \ldots, a_m)$	An ordered m-tuple.
$(a_1 a_2 \ldots a_k)$	A k-cycle; i.e. a cyclic permutation mapping $a_1 \mapsto a_2,\ a_2 \mapsto a_3,\ \ldots,\ a_{k-1} \mapsto a_k,\ a_k \mapsto a_1$.
$\binom{n}{k}$	The coefficient of x^k in the binomial expansion of $(1+x)^n$; the number of ways in which an unordered subset of k elements can be chosen from a set of n elements.

PREFACE

This book started life as a set of lecture notes and exercises for a first-year undergraduate course that I have been giving at Goldsmiths' College for several years. The idea to transform these notes into a book came initially from my students and it owes its existence in this form to their continued enthusiastic encouragement.

There is little doubt that many students find their first encounter with abstract algebra difficult. This is a pity as it may discourage them from further exploration of one of the most beautiful and satisfying areas of mathematics. One source of difficulty is that despite changes in the school syllabus over the last two decades undergraduates may still start their university course with only the haziest of notions of such fundamental topics as mappings, relations and binary operations, or with no knowledge of them at all. They can therefore be ill-equipped to plunge straight into a study of groups or rings and fields. Another problem can be their inexperience in handling the kind of proof structures so fundamental to many branches of university mathematics. Indeed they may have met very few formal proofs at school now that so little geometry is taught. Few students, in my experience, arrive at university having really grasped the logic behind the principle of finite induction or a proof by *reductio ad absurdum*.

I try to address these problems in this book both by my choice of material and by style of presentation. Chapters 1 to 5 are devoted to an introduction to the algebra of sets followed by a thorough treatment of equivalence relations, mappings, binary operations and the elementary divisibility properties of the integers. Different methods of proof are discussed and illustrated as they occur naturally in the development of these topics. Chapters 6 and 7 provide an introduction to group theory, illustrating the way in which a structure capable of great variety and wide applicability can be derived from a few simple axioms. To leave the reader with a sense of achievement, in the last section of Chapter 7 we show how it is possible to construct some finite groups of small order and the groups of order 2^m with the results developed in these chapters.

1 SETS

1.1 BASIC IDEAS

One of the most useful and fundamental ideas in mathematics is that of a *set*. By a set we simply mean a collection, list or class of objects. The objects in the set are called its *members* or *elements*. When we want to refer to a set as a whole it is convenient to refer to it by a capital letter, as, for example, the set S or A etc.; we use small letters to refer to the members of a set.

To specify a set we must describe its members and this is usually done in one of two ways:

(a) *listing* the members of the set, enclosing the list by a pair of brace brackets { };
(b) giving a *property* which distinguishes members of the set from objects not in the set.

Examples 1.1.1

1. Let A be the set of numbers 1, 2, 3, 4 and 5.

   We could write $A = \{1, 2, 3, 4, 5\}$ ('listing' method (a))

   or $A = \{x : x \text{ is an integer}, 1 \leqslant x \leqslant 5\}$ (method (b)).

   In this second notation, the brackets again denote 'is the set of . . .' and the colon after x is used to denote 'such that . . .'. So the second method of describing A reads: 'A is the set of all elements x such that x is an integer greater than or equal to 1 and less than or equal to 5.'

2. Let X be the set of solutions of the equation $x^2 - 3x + 2 = 0$.

   We write $X = \{x : x^2 - 3x + 2 = 0\}$.

1

3. Let E be the set of positive even integers.

We can adopt the 'listing' method here and write

$$E = \{2, 4, 6, 8, \ldots\}.$$

The dots continue to the end of the bracket to show that the sequence goes on indefinitely.

4. Let M be the set of r objects $m_1, m_2, \ldots, m_r$.

We write $\quad M = \{m_1, m_2, \ldots, m_r\}$

or $\quad M = \{m_i: i = 1, 2, \ldots, r\}$.

Finite and infinite sets

A set is called *finite* if it contains only a finite number of elements. Otherwise it is called *infinite*. If S is a *finite* set then the number of different elements in S is called the *order* (or *cardinality*) of S.

In examples 1.1.1, A, X and M are finite sets, while E is infinite. A has order 5 and X has order 2. We denote the order of a finite set S by $|S|$, so referring again to the examples 1.1.1 above, we could write

$$|A| = 5; \qquad |X| = 2; \qquad |M| = r.$$

(Some authors use the notation $n(A)$ to denote the order of A.)

Equality of sets

We define two sets A and B to be *equal* (write $A = B$) if and only if A and B have the same members. So to check whether $A = B$, we must check (i) whether every element of A is also an element of B and (ii) whether every element of B is also an element of A.

Example 1.1.2

Let $A = \{1, 2\}$; $B = \{x: x^2 - 3x + 2 = 0\}$; $C = \{1, 2, 2, 1, 1, 2\}$.

Here $A = B = C$. We justify saying $A = C$ by noting that

(i) both elements of A are also elements of C and
(ii) all the elements of C are also elements of A. Hence a set is determined by its *different* (or *distinct*) elements and any repetitions in the list can be ignored.

Subsets

A is said to be a *subset* of B if every element of A is an element of B.

In examples 1.1.1, A and E are both subsets of the set of integers; but A is not a subset of E since A contains elements which are not members of E.

Notation

We write '$x \in A$' to mean '(the element) x is a member of (the set) A'. Thus the symbol '$\in$' means 'belongs to'. It is used only to connect an element with a set and never between two sets. We write '$A \subseteq B$' to mean 'A is a subset of B' or '(the set) A is contained in (the set) B'. Note that this symbol is only used between two sets and never between an element and a set.

The sign $\subseteq$ can be turned the other way round: thus '$B \supseteq A$' reads '(the set) B contains (the set) A as a subset'.

We can also cancel both these signs to mean the negation of these statements; thus

'$b \notin Y$' means 'b is *not* a member of Y' and

'$S \not\subseteq Q$' means 'S is *not* a subset of Q'.

Example 1.1.3

Referring again to the sets in examples 1.1.1, the following statements are true:

(a) $4 \in A$ and $4 \in E$, but $4 \notin X$;

(b) $X = \{1, 2\}$ and hence $X \subseteq A$;

(c) $1 \in X$ but $1 \notin E$ and hence $X \not\subseteq E$;

(d) $5 \in A$ but $5 \notin E$ and hence $E \not\supseteq A$.

The empty set

It is convenient to extend our intuitive idea of a set as a collection of objects to allow for a set which has no members at all. This is called the *empty* set (or *null* set) and is denoted by the symbol $\varnothing$. A set with at least one member is called *non-empty*.

Example 1.1.4

The set $\{x: x^2 + 1 = 0, x \text{ is an integer}\}$ clearly has no members. So we can write $\{x: x^2 + 1 = 0, x \text{ is an integer}\} = \varnothing$.

Proper and improper subsets

If A is a subset of B, but $A \neq B$, we say that A is *properly* contained in B; A is called a *proper* subset of B. This relationship is written $A \subset B$.

We agree to regard the empty set $\varnothing$ as a subset of every set. Thus any non-empty set S has just two *improper* subsets, $\varnothing$ and the set S itself; all other subsets of S are *proper*.

Example 1.1.5

The set $X = \{x_1, x_2\}$ has just two proper subsets, $\{x_1\}$ and $\{x_2\}$.

The number systems

There is a standard notation for the following important sets of numbers:

$\mathbb{Z}^+ = \{1, 2, 3, \ldots\}$ denotes the set of *positive integers*.
$\mathbb{Z} = \{\ldots, -2, -1, 0, 1, 2, \ldots\}$ denotes the set of *integers*.
$\mathbb{Q} = \{m/n: m, n \in \mathbb{Z}, n \neq 0\}$ denotes the set of *rational numbers*.
$\mathbb{R}$ denotes the set of *real numbers*.
$\mathbb{C}$ denotes the set of *complex numbers*.

The choice of $\mathbb{Z}$ for the set of integers comes from the German word *zahlen*, to count. Notice that $\mathbb{Z}$ contains all the negative integers, as well as the positive ones, and that $0 \in \mathbb{Z}$. $\mathbb{Q}$ is the familiar set of fractions; the choice of $\mathbb{Q}$ comes from the word *quotient*. The real numbers are the ones that we use particularly in calculus and analysis. Notice that $\mathbb{Q}$ contains 0 and the negative rationals and $\mathbb{R}$ contains 0 and negative reals as well as the positive ones.

The axiomatic foundations and construction of these number systems are outside the scope of this book. Some properties of $\mathbb{Z}$ are developed in Chapter 4. For the moment we will use only the basic, familiar properties of the sets $\mathbb{Z}$, $\mathbb{Q}$, $\mathbb{R}$ and $\mathbb{C}$ in the illustrative examples and problems.

We can use the standard notation above to shorten our description of other sets.

Examples 1.1.6

1. The set of strictly positive real numbers = $\{x \in \mathbb{R}: x > 0\}$.
2. The set of even integers = $\{2m: m \in \mathbb{Z}\}$.
3. The set of positive integers whose last digit is $0 = \{10x: x \in \mathbb{Z}^+\}$.
4. The set of rationals with absolute value less than or equal to 1

$$= \{x \in \mathbb{Q}: -1 \leqslant x \leqslant 1\} \text{ or } \{x \in \mathbb{Q}: |x| \leqslant 1\}.$$

PROBLEMS 1.1a

1 Rewrite the following statements using set notation:
   (a) x is an element of X;   (b) a does not belong to A;
   (c) B is a subset of F;   (d) the set C is empty;
   (e) the set H contains the set D.

2 Let $A = \{a, b, c\}$. Which of the following statements are correct? Say what is wrong with each of the others.
   (a) $a \subset A$;   (b) $a \in A$;   (c) $\{a\} \subset A$;   (d) $\{a\} \in A$.

3 If $X = \{x: x^2 = 4, x > 3\}$, which of the following statements are correct? Say what is wrong with each of the others.
   (a) $X = 0$;   (b) $X = \{0\}$;   (c) $X = \varnothing$;   (d) $X = \{\varnothing\}$.

4 Describe each of the following sets by listing their elements.
   (a) $A = \{x: x > 0, x^2 = 16\}$;
   (b) $B = \{x: x > 0, x \text{ is an odd integer}\}$;
   (c) $C = \{x \in \mathbb{Z}: 0 < x < 100\}$;
   (d) $D = \{1/n: n \in \mathbb{Z}^+\}$.

5 Think of a way of describing each of the following sets by specifying a common property of the elements. (*Note*: There is no unique answer!)
   (a) $A = \{3, 4, 5, 6, 7, 8\}$;
   (b) $B = \{-\sqrt{2}, +\sqrt{2}\}$;
   (c) $C = \{1, 10, 100, 1000, \ldots\}$;
   (d) $D = \{\ldots, \frac{1}{16}, \frac{1}{8}, \frac{1}{4}, \frac{1}{2}, 1, 2, 4, 8, \ldots\}$;
   (e) $E = \{\ldots, -15, -10, -5, 0, 5, 10, \ldots\}$;
   (f) $F = \{\frac{1}{2}, \frac{2}{3}, \frac{3}{4}, \frac{4}{5}, \ldots\}$.

6 (a) Let $X = \{1, 2, 3\}$ and $Y = \{3, 1, 2\}$. Does $X = Y$?
   (b) Let A, B be any pair of sets such that $A \subseteq B$ and $B \subseteq A$. Explain why $A = B$.

5

7 Let A, B be any pair of sets such that A is a proper subset of B. Complete the following sentences in an appropriate way:

(a) Every element of A is ....

(b) At least one element of B is not ....

PROBLEMS 1.1b

1 Describe each of the following sets by listing their elements.

(a) $\{x \in \mathbb{Z}: -1 \leqslant x < 5\}$;   (b) $\{m^2: m \in \mathbb{Z}\}$;

(c) $\{n \in \mathbb{Z}^+: n < 10\}$;   (d) $\{n \in \mathbb{Z}: n < 2\}$;

(e) $\{p/q: q = 2p + 1, p \in \mathbb{Z}^+\}$.

2 Find a concise way of describing each of the following sets by specifying a common property of the elements.

(a) $\{-9, -7, -5, -3, -1, 1\}$;

(b) $\{2, 5, 10, 17, 26, 37, 50, 65, 82\}$;

(c) $\{\ldots, -8, 4, -2, 1, -\frac{1}{2}, \frac{1}{4}, -\frac{1}{8}, \pm \ldots\}$;

(d) $\{\ldots, -9, -6, -3, 0, 3, 6, 9, 12, \ldots\}$;

(e) $\{\frac{1}{2}, \frac{1}{4}, \frac{1}{6}, \frac{1}{8}, \frac{1}{10}, \ldots\}$.

3 (a) Suppose A, B and C are sets such that $A \subseteq B$ and $B \subseteq C$. Explain why $A \subseteq C$.

(b) Suppose $A \subseteq B$ and $B \subset C$ (i.e. B is a *proper* subset of C). Explain why $A \subset C$, but $A \neq C$. (*Hint*: See problem 1.1a, 7.)

1.2 SETS OF SETS

We will often be considering collections of subsets of some larger underlying set which is called the *universal* set and denoted by $\mathscr{E}$. Frequently the context makes it clear what the universal set might be and it is not necessary to describe it explicitly. For example, suppose that A denotes the set of students taking a mathematics course at your college. If we then go on to define B as the set of students taking a physics course and C as the set taking a French course etc., then it is clear that a convenient universal set would be the set of all students at your college. However, if we had defined B to be the set of students taking a mathematics course at some other college or university in the United Kingdom, then the universal set would be the set of all students taking a mathematics course at some college or university in the United Kingdom. In the case of sets of numbers, the universal set is usually clear and frequently one of the basic sets $\mathbb{Z}^+$, $\mathbb{Z}$, $\mathbb{Q}$, $\mathbb{R}$ or $\mathbb{C}$.

The complement of a set

Suppose we have a subset A of a (clearly defined) universal set $\mathscr{E}$. Then the set of elements in $\mathscr{E}$ which are *not* in A called the *complement of A in $\mathscr{E}$* and denoted by A'. Thus A' is just the set of all objects of the type we are talking about which are not members of A.

Example 1.2.1

Let $\mathscr{E} = \mathbb{Z}$ and $A = \{2n: n \in \mathbb{Z}\}$, i.e. the set of all *even* integers. Then A' is the set of all *odd* integers.

Since we are regarding $\varnothing$, the empty set, as a subset of every set, we have $\varnothing \subseteq \mathscr{E}$. It is easy to see that:

> RESULT 1.2.1 $\varnothing' = \mathscr{E}$ and $\mathscr{E}' = \varnothing$.

Power sets

The set of *all* subsets of a given set S is called the *power set* of S and denoted by $\mathscr{P}(S)$.

Example 1.2.2

Let $X = \{x_1, x_2\}$. Then the subsets of X are

  (a) with no elements: $\varnothing$;
  (b) with one element: $\{x_1\}$, $\{x_2\}$ (subsets containing just one element are sometimes called *singletons*);
  (c) with two elements: $\{x_1, x_2\} = X$, itself.

Thus $\mathscr{P}(X) = \{\varnothing, \{x_1\}, \{x_2\}, X\}$.

Note that $\mathscr{P}(X)$ is a set whose elements are themselves sets, and that the two improper subsets $\varnothing$ and X are always members of $\mathscr{P}(X)$.

PROBLEMS 1.2.1a

1 Let $S = \{1, 2, 3, 4, 5\}$ and let $A = \{1, 2\}$. Give an example of a non-empty proper subset of S satisfying each of the following conditions:

(a) a subset B such that $B \subset A$;
(b) a subset C such that $A \subset C$;
(c) the subset A';
(d) a subset E such that $E \not\subseteq A$ and $E \not\subseteq A'$.

2 Given any two real numbers x and y, then one of the two conditions '$x \leqslant y$' and '$x > y$' must be true. Taking the sets S and A in the previous problem, find an example of a subset of B of S such that *neither* of the statements:

$$A \subseteq B \qquad \text{and} \qquad A \supset B \qquad \text{is true.}$$

3 Let $X = \{x_1, x_2, x_3\}$. Find $\mathcal{P}(X)$.

A formula for $|\mathcal{P}(X)|$ when X is finite

Suppose X is a finite set of r elements, say $X = \{x_1, x_2, \ldots, x_r\}$. Clearly $|\mathcal{P}(X)|$ must depend on r. To find the formula we adopt a combinatorial approach rather different from the method used to find $\mathcal{P}(X)$ in example 1.2.2 for the case when $r = 2$. We calculate the number of different choices of a subset A, say, of X. Now the number of choices of A is just the number of different ways in which you can choose the elements of A. We consider each element in turn:

(a) we can put x_1 in or leave it out (two choices);
(b) we can put x_2 in or leave it out (two possibilities for x_2 corresponding to each possibility for x_1, giving four possibilities altogether);
(c) repeating the argument with x_3 doubles the number of possibilities again, giving $2 \times 2 \times 2 = 8$ possibilities.

The method of reasoning is illustrated in the tree diagram (Fig. 1.1).

RESULT 1.2.2 Let X be a finite set with r elements. Then

$$|\mathcal{P}(X)| = 2^r$$

There are other ways of deriving and proving this formula. We will prove it by the method of mathematical induction in Chapter 4. Another way of deriving the formula is given in problem 1.2.2a below.

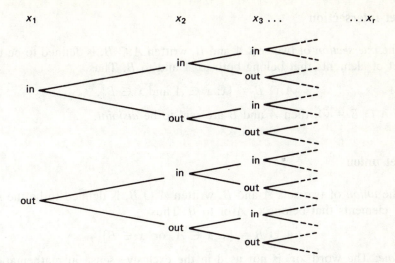

x_1 x_2 $x_3 \ldots$ $\ldots x_r$

Fig. 1.1

PROBLEM 1.2.2a

(a) Give the binomial expansion of $(1 + x)^n$, where $n \in \mathbb{Z}^+$.

(b) Deduce that $2^n = 1 + \binom{n}{1} + \binom{n}{2} + \ldots + \binom{n}{k} + \ldots + \binom{n}{n}$, where

$\binom{n}{k}$, $k = 1, 2, \ldots, n$, denotes the binomial coefficients.

(c) Let S be a finite set of n elements. By considering the number of different subsets of S containing exactly k elements,
$k = 0, 1, \ldots, n$, show that the total number of different subsets of S

is $1 + \binom{n}{1} + \binom{n}{2} + \ldots + \binom{n}{k} + \ldots + \binom{n}{n} = 2^n$.

(*Hint*: See example 1.2.2.)

1.3 COMBINING SETS

Given a pair of numbers x and y we are familiar with various ways of combining them to form a third number, $x + y$, xy or $x - y$, for example. We now define three basic ways in which a pair of sets can be combined to produce another set.

Set intersection

The *intersection* of two sets A and B, written $A \cap B$, is defined to be the set of elements that belong both to A and to B. Thus

$$A \cap B = \{x: x \in A \text{ and } x \in B\}.$$

If $A \cap B = \varnothing$, then A and B are said to be *disjoint*.

Set union

The *union* of two sets A and B, written $A \cup B$, is defined to be the set of elements that belong to A or to B. Thus

$$A \cup B = \{x: x \in A \text{ or } x \in B\}.$$

Note: The word 'or' is not used in the exclusive sense in mathematics. That is, '$x \in A$ or $x \in B$' does not exclude the possibility that x is in both A and B. Thus, if x is in either A or B or in both, then

$$x \in A \cup B.$$

Set difference

The *difference* of two sets A and B, written $A - B$, is defined to be the set of elements which are in A but not in B. The *difference* $B - A$ is similarly defined to be the set of elements which are in B but not in A. Thus

$$A - B = \{x \in A; x \notin B\}; \quad B - A = \{x \in B: x \notin A\}.$$

Venn diagrams

A useful way of illustrating sets is to use a *Venn diagram*, examples of which are shown in Fig. 1.2. The interior of the frame represents the universal set. We have the convention that the elements of each set are imagined to lie within the boundary; none lie on it.

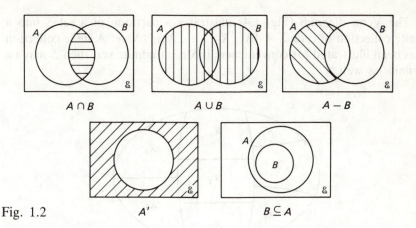

Fig. 1.2 A' $B \subseteq A$

Examples 1.3.1

1. Let A and B be the subsets of $\mathbb{Z}$ defined by:

$$A = \{1, 3, 5, 7\} \quad \text{and} \quad B = \{1, 2, 3\}.$$

(a) $A \cup B = \{1, 2, 3, 5, 7\}$;   (b) $A \cap B = \{1, 3\}$;
(c) $A - B = \{5, 7\}$;   (d) $B - A = \{2\}$.

2. Use Venn diagrams to illustrate the truth of the following results:

(a) $(A \cap B)' = A' \cup B'$;   (b) $A \cap (B \cap C) = (A \cap B) \cap C$.

Solution   We use a pair of Venn diagrams to illustrate each result, one illustrating the left-hand side of the equation and the other the right (Fig. 1.3).

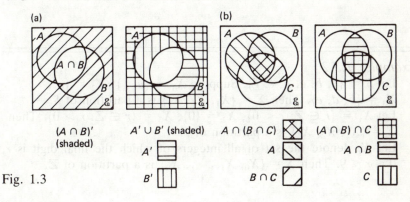

Fig. 1.3

Partitions

Let S be a set and Σ a collection of non-empty subsets of S. Then Σ is called a *partition* of S if every $x \in S$ belongs to *one* and *only one* member of Σ. We shall call the members of Σ the *cells* in the partition.

The Venn diagram (Fig. 1.4) illustrates a partition of a set S into a finite collection of *cells* $\Sigma = \{X_1, X_2, \ldots, X_8\}$. (*Note*: A *finite* collection has been illustrated for convenience; if S is an infinite set, then Σ may be infinite as well.)

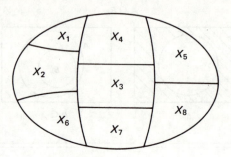

Fig. 1.4

The conditions that every element of S belongs to a member of Σ implies that the union of all the cells is the set S itself.

The condition that every member of S occurs in only one cell implies that every pair of cells is *disjoint*.

To sum up: in order to prove that a given collection Σ of subsets of a set S is a *partition* of S, we must establish three things:

> (i) Every cell $X \in \Sigma$ is non-empty.
> (ii) Every element $x \in S$ is in some cell $X \in \Sigma$.
> (iii) $X \cap Y = \varnothing$ for every pair of cells $X, Y \in \Sigma$.

Examples 1.3.2
1. Let $S = \{a, b, c, d, e, f\}$. Suppose $X_1 = \{a, f\}$, $X_2 = \{c\}$, $X_3 = \{b, d, e\}$. Then $\Sigma = \{X_1, X_2, X_3\}$ is a partition of S.
2. Let $X_1 = \{r \in \mathbb{Z}: r < 0\}$, $X_2 = \{0\}$, $X_3 = \{r \in \mathbb{Z}: r > 0\}$. Then $\Sigma = \{X_1, X_2, X_3\}$ is a partition of $\mathbb{Z}$.
3. Let X_r denote the set of all integers of which the final digit is r, $0 \leqslant r \leqslant 9$. Then $\Sigma = \{X_0, X_1, \ldots, X_9\}$ is a partition of $\mathbb{Z}$.

Counting the number of elements in a finite set

Suppose S is a finite set and Σ a partition of S. Since none of the cells are empty, it follows that Σ is a *finite* collection of cells, say

$$\Sigma = \{X_1, X_2, \ldots, X_n\}.$$

Clearly X_i is also finite, $1 \leq i \leq n$, and the condition that every element of S belongs to exactly one cell implies that

$$|S| = |X_1| + |X_2| + \ldots + |X_n|.$$

So if we can devise a partition for which we can count the number of elements in each cell then we can find $|S|$.

A formula for $|A \cup B|$

Let $S = A \cup B$, where A, B are finite sets and we are given $|A| = m_1$, $|B| = m_2$ and $|A \cap B| = x$ (Fig. 1.5).

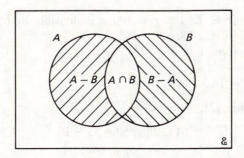

Fig. 1.5

Now if $x = 0$, A and B are disjoint and $\Sigma = \{A, B\}$ is clearly a partition of S.

Hence $$|S| = |A| + |B|. \qquad \ldots \quad (1)$$

But if $x > 0$, $\Sigma = \{A, B\}$ is not a partition of S because there is at least one element in both A and B. However,

$$\Sigma' = \{A - B, B - A, A \cap B\}$$

is a partition of S and hence

$$|S| = |A - B| + |B - A| + |A \cap B|$$
$$= (m_1 - x) + (m_2 - x) + x$$
$$= m_1 + m_2 - x.$$

This gives $$|A \cup B| = |A| + |B| - |A \cap B|. \qquad \ldots \quad (2)$$

Note that formula (1) is just a special case of (2) where $|A \cap B| = 0$. Thus formula (2) covers all possibilities.

Example 1.3.3

Find the number of integers between 1 and 100 inclusive which are divisible by 2 or by 3.

Solution Let $A = \{r \in \mathbb{Z}: 1 \leq r \leq 100, r \text{ is divisible by } 2\}$,
$B = \{r \in \mathbb{Z}: 1 \leq r \leq 100, r \text{ is divisible by } 3\}$.
Then we are required to find $|A \cup B|$.
Now $A = \{2, 4, 6, \ldots, 100\} = \{2m: 1 \leq m \leq 50\}$.
  Hence $|A| = 50$.
  $B = \{3, 6, 9, \ldots, 99\} = \{3m: 1 \leq m \leq 33\}$.
  Hence $|B| = 33$.
Now $A \cap B$ is the set of integers divisible by both 2 and 3, and hence by 6.
Thus $A \cap B = \{r \in \mathbb{Z}: 1 \leq r \leq 100, r \text{ is divisible by } 6\}$
$= \{6, 12, 18, \ldots, 96\}$
$= \{6m: 1 \leq m \leq 16\}$,

giving $|A \cap B| = 16$.

Hence by formula (2),

$$|A \cup B| = 50 + 33 - 16$$

$$= 67.$$

PROBLEMS 1.3a

1 Let A, B, C be the subsets of $\mathbb{Z}^+$ defined by
$A = \{1, 2, 3, 4, 5\}$; $B = \{3, 5, 7\}$; $C = \{2, 4, 6\}$.
Describe the following subsets:
(a) $A \cup B$;   (b) $A \cap C$;
(c) $A - B$;   (d) $B - C$;
(e) $A \cap (B \cup C)$;   (f) $(A \cap B) \cup (A \cap C)$;
(g) $A' \cap C'$;   (h) $(A \cup C)'$.

2 Let X, Y, Z be the subsets of $\mathbb{Z}^+$ defined by:
$X = \{4n: n \in \mathbb{Z}^+\}$; $Y = \{6n: n \in \mathbb{Z}^+\}$; $Z = \{5n: n \in \mathbb{Z}^+\}$.
(a) List the elements of X, Y and Z;
(b) describe the sets $X \cap Y$, $X \cap Z$ and $Y \cap Z$.

3 Use Venn diagrams to illustrate the truth of the following results:
(a) $A \cap (B \cup C) = (A \cap B) \cup (A \cap C)$;
(b) $A \cup (B \cup C) = (A \cup B) \cup C$; (c) $(A \cup B)' = A' \cap B'$.

4 Let A be a subset of $\mathscr{E}$. Complete the following:

$A \cap A =$; $A \cap A' =$; $A \cap \mathscr{E} =$; $A \cap \varnothing =$;

$A \cup A =$; $A \cup A' =$; $A \cup \mathscr{E} =$; $A \cup \varnothing =$.

5 Let A, B both be subsets of a universal set $\mathscr{E}$. Find the relation between A and B in each of the following cases. Try to give the most *general* answer possible:

(a) $A \cap B = A$; (b) $A \cup B = A$; (c) $B' \subset A'$.

6 Figure 1.6 illustrates three subsets A, B, C of a universal set $\mathscr{E}$. Using the operations of intersection and set complement only, label the remaining seven disjoint subsets.

Hence define a partition $\Sigma = \{X_1, X_2, \ldots, X_7\}$ of $A \cup B \cup C$.

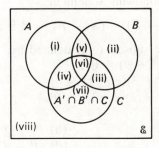

Fig. 1.6

7 Let A, B, C be finite subsets of a universal set $\mathscr{E}$. By letting $|X_i| = x_i$, $1 \leqslant i \leqslant 7$, where $\Sigma = \{X_1, X_2, \ldots, X_7\}$ is the partition of $A \cup B \cup C$ obtained in problem 6, or otherwise, prove that

$|A \cup B \cup C| =$

$|A| + |B| + |C| - |A \cap B| - |B \cap C| - |C \cap A| + |A \cap B \cap C|$.

8 112 students sat for a certain examination. Question 1 was attempted by 50 students, question 2 by 66 students and question 3 by 38 students. 32 students attempted both questions 1 and 2, 22 students attempted questions 2 and 3 and 20 students attempted questions 1 and 3.

If only 8 students attempted all the questions 1, 2 and 3, how many students attempted none of these three questions?

PROBLEMS 1.3b

1 S_1, S_2, S_3, S_4 are four sets such that $|S_i| = 4$, $i = 1, 2, 3, 4$, and $|S_i \cap S_j| = 1$, $1 \leqslant i < j \leqslant 4$. Draw Venn diagrams to show the three possible ways in which the four sets intersect. In each case show that it is possible to colour the elements of $S_1 \cup S_2 \cup S_3 \cup S_4$ using four colours so that each set contains exactly one element of each colour. (The general problem of whether it is possible to colour the elements of n sets intersecting in this way for all integers n, so that each set contains exactly one element of each colour, is still unsolved. This problem was posed by the contemporary Hungarian mathematician Paul Erdös.)

2 (a) Let X, Y, Z be subsets of a set $\mathscr{E}$. Illustrate using a Venn diagram the subsets (i) $X - (Y-Z)$ and (ii) $(X - Y) - Z$. Deduce that in general $X - (Y - Z) \neq (X - Y) - Z$.

  (b) Let $\mathscr{E} = \{1, 2, 3\}$. Give an example of three non-empty proper subsets X, Y, Z of $\mathscr{E}$ such that
(i) $X - (Y - Z) \neq (X - Y) - Z$;
(ii) $X - (Y - Z) = (X - Y) - Z$.

  (c) Complete the following statement:

$$X - (Y - Z) = (X - Y) - Z \text{ if and only if } \ldots .$$

3 Let P, Q, S be the subsets of $\mathbb{R}$ defined by:
$P = \{x: (x^2 - 4)(x^2 - 9) = 0\}$;
$Q = \{x: (x - 2)(x + 3)(x - 1) = 0\}$;
$S = \{x: (x^2 - 1)(x + 2)(x - 3) = 0\}$.
Describe the sets
(a) $P \cup Q \cup S$;   (b) $P \cap Q \cap S$;
(c) $P - (Q - S)$;   (d) $(P - Q) - S$.

4 The *symmetric difference* of two sets A and B, written $A \bigtriangleup B$, is defined to be the set $(A \cup B) - (A \cap B)$. Illustrate this set using a Venn diagram and show that $A \bigtriangleup B = (A - B) \cup (B - A)$.
Describe the sets
(a) $A \bigtriangleup A$;   (b) $A \bigtriangleup \varnothing$;
(c) $A \bigtriangleup \mathscr{E}$ where $A \subset \mathscr{E}$;   (d) $A \bigtriangleup B$ where $A \supset B$.

5 A market research organisation finds that in a certain district 90 households claimed to use at least one of the three products P, Q and R. 50 claimed to use P, 71 to use Q and 37 to use R. If only 9 households claimed to use all three products, show that
(a) at least 31 claimed to use both P and Q;
(b) at least 18 claimed to use both Q and R;
(c) at most 28 claimed to use both P and R.

6 Let A, B, C be subsets of $\mathscr{E}$. Illustrate with a Venn diagram the sets
(a) $A \cap (B' \cap C)'$;   (b) $B \cup (A \cap C')$.
Does the diagram suggest a containment relation between these two sets?

7 Let A, B, C be finite subsets of some universal set $\mathscr{E}$. Let

$$D = B \cup C.$$

By substituting for D in the formula

$$|A \cup D| = |A| + |D| - |A \cap D|$$

and making repeated use of the formula for $|X \cup Y|$, prove the formula for $|A \cup B \cup C|$ (see problem 1.3a, 7).
Note: You will also need the identity

$$A \cap (B \cup C) = (A \cap B) \cup (A \cap C)$$

(see problem 1.3a, 3(a)).

1.4 SETS AND LOGIC

Venn diagrams can be used to prove results and identities when combinations of only two or three sets are involved, because it is possible to draw a diagram illustrating the most general possible case (see, for example, examples 1.3.1, 2(a) and (b)). For more than three sets this becomes very difficult and a reasoning proof is therefore necessary. We first introduce a symbol which is useful in all logical and deductive proofs.

Logical implication

The symbol $\Rightarrow$ reads 'implies that'. It is used to connect two statements or conditions.

Example 1.4.1

(a) It is snowing $\Rightarrow$ the weather is cold.
(b) n is an integer $\Rightarrow$ n is a real number.
(c) n is an integer divisible by 5 $\Rightarrow$ n is an integer whose final digit is 0 or 5.
(d) Let $x \in \mathbb{R}$; $x \geqslant 7 \Rightarrow x > 3$.
(e) Let $m \in \mathbb{Z}$; $m \geqslant 7 \Rightarrow m > 6$.

In example 1.4.1, sentences such as 'it is snowing', 'the weather is cold' and '$m \geqslant 7$ implies that $m > 6$' are called *statements*. They are either true or false. Sentences such as 'n is an integer', '$x > 3$' etc. are called *conditions* on the *variable n* (or x). A condition may be true for some values of the variable and false for others. For example, '$x > 3$' is true when $x = 10$ but false when $x = 1$.

We have said that the statement

'Condition p holds $\Rightarrow$ condition q holds'

17

can be read

'Condition p holds *implies that* condition q holds.'

The following are alternative wordings of the *same* statement:

(a) If condition p holds, *then* condition q holds.
(b) Condition q *is implied by* condition p.
(c) Condition q holds *if* condition p holds.
(d) p is a *sufficient* condition for q.

Note, also, that if q does *not* hold, then p cannot hold so that the following are also alternative readings of the statement 'condition p holds $\Rightarrow$ condition q holds'.

(e) Condition p holds *only if* condition q holds.
(f) q is a *necessary* condition for p.

As with other symbols we have introduced, the implication sign can be written the other way round.

'Condition p holds $\Leftarrow$ condition q holds'

can be read

'Condition p holding *is implied by* condition q holding.'

We can obtain six alternative and more natural ways of wording this statement by interchanging p and q in (a) to (f) above.

Equivalent conditions

It may happen that given two conditions p and q, both the statements

'p holds $\Rightarrow$ q holds'   and   'q holds $\Rightarrow$ p holds'

are true. In this case the two statements are contracted symbolically into one and we write

'p holds $\Leftrightarrow$ q holds'

This is usually read

'p holds if and only if q holds'   or   'p is a necessary and sufficient condition for q'

and we therefore call p and q *equivalent* conditions.

Let P, Q be the set of values of the variable for which the conditions p and q hold respectively. Then the statement

'p holds $\Rightarrow$ q holds'

can be expressed in terms of the sets P and Q as

$$P \subseteq Q.$$

Similarly, 'q holds $\Rightarrow p$ holds' is the same as saying '$Q \subseteq P$'.

Thus 'p holds $\Leftrightarrow q$ holds' can be expressed as $P = Q$.

Example 1.4.2

Referring to example 1.4.1 above,

  (a) It is snowing $\not\Leftrightarrow$ the weather is cold. These are clearly *not* equivalent statements.
  (b) n is an integer $\not\Leftrightarrow n$ is a real number. These conditions are not equivalent. Similarly, the conditions in (d) are not equivalent.
  (c) However, n is an integer divisible by 5 $\Leftrightarrow n$ is an integer whose final digit is 0 or 5.
  Hence these conditions are equivalent.
  The conditions in (e) are also equivalent.

Example 1.4.3

Let A, B be subsets of a set. We rewrite some of the definitions given in the previous sections using the implications signs. Let $x \in \mathscr{E}$. Then

  (a) $A \subseteq B$ if $(x \in A \Rightarrow x \in B)$; (b) $A = B$ if $(x \in A \Leftrightarrow x \in B)$;
  (c) $x \in A \cap B \Leftrightarrow (x \in A$ and $x \in B)$;
  (d) $x \in A \cup B \Leftrightarrow (x \in A$ or $x \in B)$;
  (e) $x \in A - B \Leftrightarrow (x \in A$ and $x \notin B)$;
  (f) $x \in A' \Leftrightarrow x \notin A$.

In the following examples we show how to utilise the symbolic statements in example 1.4.3 to build up a proof of a more complicated result or identity.

Examples 1.4.4

1. Prove that if $A \subseteq B$ and $A' \subseteq B'$, then $A = B$.
   (To prove $A = B$, we must show $A \subseteq B$ and $B \subseteq A$. We are given $A \subseteq B$, so it remains only to show $B \subseteq A$. To do this we must show $x \in B \Rightarrow x \in A$.)

Solution $x \in B \Rightarrow x \notin B'$

$\Rightarrow x \notin A'$ (since $A' \subseteq B'$)

$\Rightarrow x \in A$.

Hence $B \subseteq A$, and, since $A \subseteq B$ (given), $A = B$.

2. Prove that if A and B are subsets of set X, then $A \cup B \subseteq X$. (We need to show $x \in A \cup B \Rightarrow x \in X$.)

Solution $x \in A \cup B \Rightarrow x \in A$ or $x \in B$.

But $x \in A \Rightarrow x \in X$, since $A \subseteq X$ (given).

Similarly, $x \in B \Rightarrow x \in X$, since $B \subseteq X$ (given).

Thus $x \in A \cup B \Rightarrow x \in X$.

Hence $A \cup B \subseteq X$.

3. Prove that $A \cap (B \cup C) = (A \cap B) \cup (A \cap C)$.

(By example 1.4.3 (b), to prove $X = Y$ we must show $x \in X \Leftrightarrow x \in Y$.)

Solution $x \in A \cap (B \cup C) \Leftrightarrow x \in A$ and $x \in B \cup C$

$\Leftrightarrow x \in A$ and $x \in (B$ or $C)$

$\Leftrightarrow (x \in A$ and $x \in B)$ or $(x \in A$ and $x \in C)$

$\Leftrightarrow x \in (A$ and $B)$ or $x \in (A$ and $C)$

$\Leftrightarrow x \in A \cap B$ or $x \in A \cap C$.

$\Leftrightarrow x \in (A \cap B) \cup (A \cap C)$.

Hence $A \cap (B \cup C) = (A \cap B) \cup (A \cap C)$.

De Morgan's laws

The following two identities, involving set complements, are known as De Morgan's laws:

(I) $(A \cap B)' = A' \cup B'$;   (II) $(A \cup B)' = A' \cap B'$.

Before giving a formal proof, we illustrate these identities in Fig. 1.7.

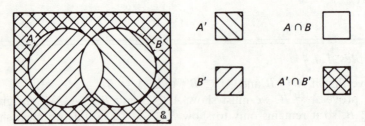

Fig. 1.7

Proof of (I) $x \in (A \cap B)'$

$\Leftrightarrow x \notin A \cap B$

$\Leftrightarrow x \notin (A$ and $B)$

$\Leftrightarrow x \notin A$ or $x \notin B$ (since x is not in both A and B, then x must be missing from A or from B).

$\Leftrightarrow x \in A'$ or $x \in B'$

$\Leftrightarrow x \in A' \cup B'$.

Notice particularly the point in the proof at which we change from 'and' to 'or'.

The proof of (II) is similar and left as an exercise (problem 1.4a, 5).

PROBLEMS 1.4a

1 Put the correct sign $\Rightarrow$, $\Leftarrow$ or $\Leftrightarrow$ between the following pairs of conditions on a real number x.

(a) $x \in \mathbb{Q}$; the representation of x as a decimal terminates after a finite number of digits.

(b) $x > 5$; $x \geqslant 6$.

(c) $x \in \mathbb{Q}$; $\dfrac{1}{x} \in \mathbb{Q}$.

(d) $x \in \mathbb{Z}$; $1 - x \in \mathbb{Z}$.

In each of the following problems A, B, C are subsets of a set $\mathscr{E}$.

2 Prove that $A \cap B \subseteq A$.

3 Prove that $A \cap B = \varnothing \Rightarrow A \subseteq B'$. Is $B \subseteq A'$ also true?

4 Prove that $A - B = A \cap B'$.

5 Prove that $A' \cap B' = (A \cup B)'$.
Hence show that $A' \cap (B \cup C)' = (A \cup B)' \cap C'$.

6 Prove that
(a) $(A')' = A$;   (b) $A \subseteq (A' \cap B)'$.
Hence prove that $(A' \cap B)' \cup C = A \cup C$ if and only if $B' \subseteq A \cup C$.
(You may assume De Morgan's law.)

7 Prove that
$(A \cap B) \cup (B \cap C) \cup (C \cap A) = (A \cup B) \cap (B \cup C) \cap (C \cup A)$.
What identity is obtained by putting $C = \varnothing$ in this result?

PROBLEMS 1.4b

In the following problems, A, B, C, D are subsets of a set $\mathscr{E}$.

1 Put the correct sign $\Rightarrow$, $\Leftarrow$ or $\Leftrightarrow$ between the following pairs of conditions on a real number x:

(a) $x \in \mathbb{Z}^+; \dfrac{1}{x} \in \mathbb{Q}$. $\qquad$ (b) $3x \in \mathbb{Z}; x + 1 \in \mathbb{Z}$.

(c) $x \in \mathbb{Q}; 2x \in \mathbb{Q}$. $\qquad$ (d) $x \in \mathbb{Q}; \dfrac{1}{1-x} \in \mathbb{Q}$.

2 Prove that $A \cup (B \cap C) = (A \cup B) \cap (A \cup C)$.

3 Prove that $A' - B' = B - A$.

4 Prove that $(A - B) \cap (A - C) = A - (B \cup C)$.
  Find and prove a similar identity for $(A - B) \cup (A - C)$.

5 Prove that $A - (A - B) = A \cap B$.

6 Let $P = (A \cap B) \cup (C \cap D)$; $Q = (A \cup C) \cap (B \cup D)$.
  Prove that $P \subseteq Q$.
  Let $\mathscr{E} = \mathbb{Z}^+$, $A = \{1\}$ and $B = \{2\}$. Show that it is possible to define the subsets C and D of $\mathbb{Z}^+$ so that $P = \varnothing$ and $Q \neq \varnothing$. Deduce that $Q \not\subseteq P$.

7 Prove that $A \cap (B' \cap C)' \subseteq B \cup (A \cap C')$ and that equality holds if and only if $B \cap A' = \varnothing$.

8 Prove that $A - (B - C) \supseteq (A - B) - C$. State and prove a necessary and sufficient condition for equality to hold.
  (*Hint*: See problem 1.3b, 2; but a reasoning proof is required here.)

1.5 CARTESIAN PRODUCT SETS

We are familiar with the idea of a point in plane coordinate geometry being represented by a pair of coordinates (x, y). Furthermore, the point with coordinates (x, y) is different from the point with coordinates (y, x) unless $x = y$. The coordinates (x, y) are an example of an *ordered* pair, so called since the *order* in which x and y appear inside the bracket matters.

Equality of ordered pairs

Two ordered pairs are defined to be equal if and only if their first coordinates are equal and their second coordinates are equal. That is,

$$(x_1, y_1) = (x_2, y_2) \Leftrightarrow x_1 = x_2 \text{ and } y_1 = y_2.$$

Cartesian product set

Let A, B be sets. Then the set of all *ordered* pairs (a, b) with $a \in A$ and $b \in B$ is called the *cartesian product* of A by B, denoted by $A \times B$ (and read 'A cross B'). That is,

$$A \times B = \{(a, b): a \in A \text{ and } b \in B\}.$$

Examples 1.5.1

1. Let $A = \{1, 2\}$ and $B = \{p, q\}$.

   Then $\qquad A \times B = \{(1, p), (1, q), (2, p), (2, q)\}$,

   and $\qquad B \times A = \{(p, 1), (p, 2), (q, 1), (q, 2)\}$.

2. In defining $A \times B$ we did not exclude the possibility that A or B is empty. But it is clear that for any set A, $A \times \varnothing = \varnothing$, since we have no choice for a second coordinate. Similarly $\varnothing \times A = \varnothing$.

Cartesian diagrams

We can illustrate $A \times B$ by drawing a cartesian coordinate diagram.

Example 1.5.2

Let $A = \{1, 2, 3\}$ and $B = \{1, 2\}$. Then the elements of $A \times B$ can be illustrated by the points in the diagram (Fig. 1.8).

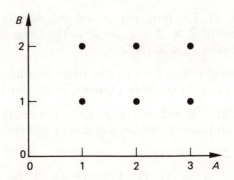

Fig. 1.8

Ordered n-tuples

We can generalise the idea of an ordered pair to an ordered sequence of n-elements, for any $n \in \mathbb{Z}^+$. We call such a sequence $(a_1, a_2, \ldots, a_n)$ an *ordered n-tuple*.

Equality of ordered n-tuples

We define

$$(a_1, a_2, \ldots, a_n) = (b_1, b_2, \ldots, b_n) \Leftrightarrow a_i = b_i, \quad i = 1, 2, \ldots, n.$$

Notice that no equality can exist between an ordered n-tuple and an ordered m-tuple unless $n=m$.

Cartesian product of several sets

Let $A_1, A_2, \ldots, A_n$ be any n sets, $n \in \mathbb{Z}^+$. Then the set of all ordered n-tuples $(a_1, a_2, \ldots, a_n)$, where $a_i \in A_i$, $i = 1, 2, \ldots, n$, is called the *cartesian product* $A_1 \times A_2 \times \ldots \times A_n$. That is

$$A_1 \times A_2 \times \ldots \times A_n = \{(a_1, a_2, \ldots, a_n): a_i \in A_i, i = 1, 2, \ldots, n\}.$$

If $A_1 = A_2 = \ldots A_n$, then the cartesian product $A_1 \times A_2 \times \ldots A_n$ is often denoted by A^n. In particular, $A \times A$ is frequently written A^2.

Examples 1.5.3

1. $(1, 2, 0) \neq (1, 2)$. The former is an *ordered triple* (or *ordered 3-tuple*) and an element of $\mathbb{Z} \times \mathbb{Z} \times \mathbb{Z}$ (or $\mathbb{Z}^3$). The latter is an *ordered pair* and an element of $\mathbb{Z} \times \mathbb{Z}$ (or $\mathbb{Z}^2$).

2. An ordered n-tuple may have two or more coordinates the same: $(1, 1, 1)$ and $(1, 2, 1)$ are both elements of $\mathbb{Z}^3$, for example.

3. The set $\mathbb{R}^3 = \mathbb{R} \times \mathbb{R} \times \mathbb{R} = \{(x, y, z): x, y, z \in \mathbb{R}\}$ can be regarded as the set of all three-dimensional cartesian coordinates.

4. Let $A = \{1, 2\}$.

   Then $\qquad A^2 = \{(1, 1), (1, 2), (2, 1), (2, 2)\}.$

PROBLEMS 1.5a

1 For each of the following pairs of sets A and B, find $A \times B$, $B \times A$ and $(A \times B) \cap (B \times A)$.
(a) $A = \{1, 2\}$, $B = \{5, 6\}$.   (b) $A = \{1, 2\}$, $B = \{1, 3\}$.
(c) $A = \{1, 2\}$, $B = \{1, 2, 3\}$.

2 Let A, B be any pair of non-empty sets. Prove that
$(A \times B) \cap (B \times A) = (A \cap B) \times (A \cap B)$.

3 Suppose $A \times B$ is non-empty. Prove that $A \times B = C \times D \Rightarrow A = C$ and $B = D$. Why is this not necessarily true if $A \times B = C \times D = \varnothing$?

4 Suppose A, B are finite sets with $|A| = m$ and $|B| = n$. State and prove a formula for $|A \times B|$.

5 Mark on a coordinate diagram the following subsets of $\mathbb{R} \times \mathbb{R}$:
(a) $\{(x, y): x = 0\}$;   (b) $\{(x, y): x > y\}$,
(c) $\{(x, y): x + y = 1\}$;   (d) $\{(x, y): x^2 + y^2 \leq 4\}$.

6 Let $A = \{1, 2\}$, $B = \{3, 4\}$, $C = \{4, 5\}$. Find
(a) $A \times (B \cap C)$;   (b) $(A \times B) \cap (A \times C)$;
(c) $A \times (B \cup C)$;   (d) $(A \times B) \cup (A \times C)$.

7 For any sets A, B, C prove that
(a) $A \times (B \cap C) = (A \times B) \cap (A \times C)$;
(b) $A \times (B \cup C) = (A \times B) \cup (A \times C)$.

PROBLEMS 1.5b

1 Let $X = \{0, 1\}$. List the elements of X^3.

2 Let A be a finite set with $r > 0$ elements. State and prove a formula for $|A^n|$, $n \in \mathbb{Z}^+$.

3 Shade on a coordinate diagram the sets
(a) $\{(x, y): 0 \leq x \leq 1, 0 \leq y \leq 1\} \cap \{(x, y): y > x^2\}$;
(b) $\{(x, y): x > 2y\} \cup \{(x, y): x + y \leq 2\}$.

4 Let $A = \{1, 2\}$, $B = \{2, 3\}$, $C = \{3, 4\}$, $D = \{4, 5\}$.
Find the sets
(a) $(A \cap B) \times (C \cap D)$;   (b) $(A \times C) \cap (B \times D)$.

Prove that for any sets A, B, C, D,

$$(A \times C) \cap (B \times D) = (A \cap B) \times (C \cap D).$$

5 For the sets $A = \{1\}$, $B = \{2\}$, $C = \{3\}$, $D = \{4\}$, find
(a) $(A \times C) \cup (B \times D)$;   (b) $(A \cup B) \times (C \cup D)$.
Prove that for any sets A, B, C, D,

$$(A \times C) \cup (B \times D) \subseteq (A \cup B) \times (C \cup D).$$

Deduce that equality does not always hold and find and prove a formula for $(A \cup B) \times (C \cup D)$.

SOLUTIONS TO PROBLEMS IN THE (a) SETS

Problems 1.1a

1 (a) $x \in X$;   (b) $a \notin A$;   (c) $B \subseteq F$;   (d) $C = \emptyset$;
(e) $H \supseteq D$.

2 Correct statements: (b) and (c); (a) is wrong because a is an *element* and $\subset$ is only used between *sets*; (d) is wrong because $\{a\}$ denotes the *subset* containing a and $\in$ is only used between an element and a set.

3 $X = \emptyset$ is correct; (a) 0 is not used to denote the empty set; (b) $\{0\}$ denotes the set containing 0 as an element, but $0 \notin X$; (d) $\{\emptyset\}$ denotes a set containing the empty set as its only element, and thus is not empty!

4 (a) $A = \{4\}$;   (b) $B = \{1, 3, 5, 7, \ldots\}$;
(c) $C = \{1, 2, 3, \ldots, 99\}$;   (d) $D = \{1, \frac{1}{2}, \frac{1}{3}, \frac{1}{4}, \ldots\}$.

5 (a) $A = \{n \in \mathbb{Z}: 2 < n < 9\}$;   (b) $B = \{x: x^2 = 2\}$;
(c) $C = \{10^r: r = 0, 1, 2, \ldots\}$;   (d) $D = \{2^r: r \in \mathbb{Z}\}$;
(e) $E = \{5m: m \in \mathbb{Z}\}$;   (f) $F = \left\{\dfrac{n}{n+1}: n \in \mathbb{Z}^+\right\}$.

6 (a) Yes; (b) $A \subseteq B$ implies that every element of A is also in B; but $B \subseteq A$ implies that every element of B is also in A; hence A and B have the same members.

7 (a) ... an element of B;   (b) ... an element of A.

Problems 1.2.1a

1 (a) $B = \{1\}$ or $B = \{2\}$;
(b) $C = \{1, 2, 3\}$ is one possibility (there are six possibilities for C in all);
(c) $A' = \{3, 4, 5\}$;
(d) $E = \{2, 3, 5\}$ is one possibility (E must contain at least one element of A and one element of A').

2 $B = \{1, 3\}$ is an example (B fits the conditions if B contains 0 or 1 element of A and 1, 2 or 3 elements of A').

3 $\mathcal{P}(X) = \{\emptyset, \{x_1\}, \{x_2\}, \{x_3\}, \{x_2, x_3\}, \{x_1, x_3\}, \{x_1, x_2\}, \{x_1, x_2, x_3\}\}$.

Problems 1.2.2a

(a) $(1 + x)^n = 1 + \binom{n}{1}x + \binom{n}{2}x^2 + \ldots + \binom{n}{k}x^k + \ldots + \binom{n}{n}x^n$.

(b) Put $x = 1$ in (d);

(c) there are $\binom{n}{k}$ ways of choosing a subset of k elements from a set of n elements, since the order of the elements does not matter. There is just one subset of S containing no elements, that is $\emptyset$. Hence there are

$$1 + \sum_{k=1}^{n} \binom{n}{k}$$

subsets of S altogether. Using (b) gives the result.

Problems 1.3a

1 (a) $\{1, 2, 3, 4, 5, 7\}$;   (b) $\{2, 4\}$;
(c) $\{1, 2, 4\}$;   (d) $\{3, 5, 7\}$;
(e) $\{2, 3, 4, 5\}$;   (f) $\{2, 3, 4, 5\}$;
(g) $\{7, 8, 9, \ldots\}$;   (h) $\{7, 8, 9, \ldots\}$.

2 (a) $X = \{4, 8, 12, 16, \ldots\}$; $Y = \{6, 12, 18, 24, \ldots\}$;
$Z = \{5, 10, 15, 20, \ldots\}$.
(b) $X \cap Y = \{12n: n \in \mathbb{Z}^+\}$; $X \cap Z = \{20n: n \in \mathbb{Z}^+\}$;
$Y \cap Z = \{30n: n \in \mathbb{Z}^+\}$.

3 See Fig. 1.9.

4 A; $\emptyset$; A; $\emptyset$;
A; $\mathscr{E}$; $\mathscr{E}$; A.

5 (a) $A \subseteq B$; (b) $B \subseteq A$; (c) $A \subset B$.

6 (i) $A \cap B' \cap C'$;   (ii) $B \cap A' \cap C'$;
(iii) $A' \cap B \cap C$;   (iv) $A \cap B' \cap C$;
(v) $A \cap B \cap C'$;   (vi) $A \cap B \cap C$;
(viii) $A' \cap B' \cap C'$.

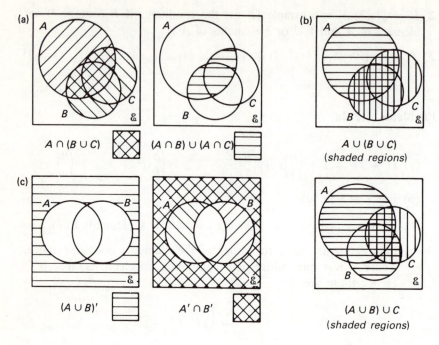

Fig. 1.9

In Σ, let $X_1, X_2, \ldots, X_7$ be given by subsets defined in (i) to (vii).

7 With the notation of problem 6, using the diagram for problem 6 gives:

$|A| = x_1 + x_4 + x_5 + x_6; \quad |B| = x_2 + x_3 + x_5 + x_6;$
$|C| = x_3 + x_4 + x_6 + x_7.$

$|A \cap B| = x_5 + x_6$ etc. and substituting in right-hand side of the given formula gives

$|A \cup B \cup C| = x_1 + x_2 + \ldots + x_7 = |X_1| + |X_2| + \ldots + |X_7|,$
as required.

8 Let A, B, C be the set of students who attempted questions 1, 2 and 3 respectively. Then we are given:

$|A| = 50, |B| = 66, |C| = 38; |A \cap B| = 32;$
$|B \cap C| = 22; |C \cap A| = 20; |A \cap B \cap C| = 8.$

The number of students who attempted at least one of questions 1, 2 and 3 = $|A \cup B \cup C|$ = $50 + 66 + 38 - 32 - 22 - 20 + 8$ (see problem 7) = 88.

$\therefore$ Number attempting none of questions 1, 2 or 3 = $112 - 88 = 24$.

Problems 1.4a

1 (a) $\Leftarrow$ (b) $\Leftarrow$ (c) $\Leftarrow$ (note that $x \in \mathbb{Q}$ includes the possibility
$x = 0$; $\dfrac{1}{x} \notin \mathbb{Q}$ if $x = 0$)   (d) $\Leftrightarrow$.

2 $x \in A \cap B \Rightarrow x \in A$ and $x \in B$, so every element of $A \cap B$ is in A.
Hence $A \cap B \subseteq A$.

3 Suppose $A \cap B = \varnothing$. Then $x \in A \Rightarrow x \notin B \Rightarrow x \in B' \Rightarrow A \subseteq B'$.
Yes (reverse the roles of A and B).

4 $x \in A - B \Rightarrow x \in A$ and $x \notin B \Rightarrow x \in A$ and $x \in B'$
$\Rightarrow x \in A \cap B'$.

5 $x \in A' \cap B' \Leftrightarrow x \in A'$ and $x \in B' \Leftrightarrow x \notin A$ and $x \notin B$
$\Leftrightarrow x \notin (A \text{ or } B) \Leftrightarrow x \notin A \cup B \Leftrightarrow x \in (A \cup B)'$.
Hence $x \in A' \cap (B \cup C)' \Leftrightarrow x \in A'$ and $x \in (B \cup C)'$
$\Leftrightarrow x \in A'$ and $x \in B' \cap C' \Leftrightarrow x \in (A' \text{ and } B' \text{ and } C')$
$\Leftrightarrow x \in (A' \cap B')$ and $x \in C' \Leftrightarrow x \in (A \cup B)'$ and $x \in C'$
$\Leftrightarrow x \in (A \cup B') \cap C'$.

6 (a) $x \in (A')' \Leftrightarrow x \notin A' \Leftrightarrow x \in A$. Hence $(A')' = A$.
(b) Using De Morgan's law, $(A' \cap B)' = (A')' \cup B' = A \cup B'$.
Hence $x \in A \Rightarrow x \in (A \text{ or } B') \Rightarrow x \in A \cup B' \Rightarrow x \in (A' \cap B)'$,
so $A \subseteq (A' \cap B)'$.
Denote the set $(A' \cap B)' \cup C$ by X. Now $x \in A \cup C \Rightarrow x \in A$ or
$x \in C$. But $x \in A \Rightarrow x \in (A' \cap B)'$. Hence $x \in A$ or $x \in C$
$\Rightarrow x \in (A' \cap B)'$ or $x \in C \Rightarrow x \in X$. Hence $A \cup C \subseteq X$, and this is
true for any set B. Thus we need to show that $B' \subseteq A \cup C \Leftrightarrow$
$X \subseteq A \cup C$.
Suppose first that $B' \subseteq A \cup C$.
Then $x \in X \Rightarrow x \in (A' \cap B)'$ or $x \in C \Rightarrow x \in A \cup B'$ or
$x \in C \Rightarrow x \in (A \text{ or } B')$ or $x \in C \Rightarrow x \in (A \text{ or } B' \text{ or } C)$. But by
hypothesis $B' \subseteq A \cup C$, and hence $x \in B' \Rightarrow x \in (A \text{ or } C)$. Hence
$x \in X \Rightarrow x \in (A \text{ or } C)$
Thus $X \subseteq A \cup C$ and this together with $A \cup C \subseteq X$ proves that
$B' \subseteq A \cup C \Rightarrow X = A \cup C$.
Finally, suppose $X = A \cup C$. Then $x \in X \Rightarrow x \in A \cup C$. But we have
shown above that $x \in X \Leftrightarrow x \in (A \text{ or } B' \text{ or } C)$. Hence
$x \in (A \text{ or } B' \text{ or } C) \Rightarrow x \in A$ or C.
Thus in particular, $x \in B' \Rightarrow x \in (A \text{ or } C) \Rightarrow x \in A \cup C$.
Thus $X = A \cup C \Rightarrow B' \subseteq A \cup C$, completing the proof.

7 Denote $(A \cap B) \cup (B \cap C) \cup (C \cap A)$ by X
and $(A \cup B) \cap (B \cup C) \cap (C \cup A)$ by Y.
Then $x \in X \Rightarrow x \in (A \text{ and } B)$ or $x \in (B \text{ and } C)$ or $x \in (C \text{ and } A)$.

Now $x \in A \Rightarrow x \in (A \cup B$ and $C \cup A)$.

$x \in B \Rightarrow x \in (A \cup B$ and $B \cup C)$.

$x \in C \Rightarrow x \in (B \cup C$ and $C \cup A)$.

Hence $x \in (A$ and $B) \Rightarrow x \in A \cup B$ and $x \in B \cup C$ and $x \in C \cup A$.

Thus $x \in (A$ and $B) \Rightarrow x \in Y$. Similarly $x \in (B$ and $C) \Rightarrow x \in Y$ and $x \in (C$ and $A) \Rightarrow x \in Y$. Thus $x \in X \Rightarrow x \in Y$ and $X \subseteq Y$.

Next suppose that $y \in Y$. Then $y \in (A$ or $B)$ and $y \in (B$ or $C)$ and $y \in (C$ or $A)$. Now either $y \in A$ or $y \notin A$. If $y \notin A$, then $y \in B$ and $y \in C$ and hence $y \in B \cap C$ and so $y \in X$. If $y \in A$, then since $y \in (B$ or $C)$ we have $y \in (A$ and $B)$ or $y \in (A$ and $C)$. In either case $y \in X$. Thus $y \in Y \Rightarrow y \in X$.

Hence $Y \subseteq X$ and this with $X \subseteq Y$ proves $X = Y$.

Putting $C = \varnothing$ gives: $A \cap B = (A \cup B) \cap B \cap A$.

Problems 1.5a

1 (a) $A \times B = \{(1, 5), (1, 6), (2, 5), (2,6)\}$;

   $B \times A = \{(5, 1), (6, 1), (5, 2), (6, 2)\}$;

   $(A \times B) \cap (B \times A) = \varnothing$.

   (b) $A \times B = \{(1, 1), (1, 3), (2, 1), (2, 3)\}$;

   $B \times A = \{(1, 1), (3, 1), (1, 2), (3, 2)\}$;

   $(A \times B) \cap (B \times A) = \{(1, 1)\}$.

   (c) $A \times B = \{(1, 1), (1, 2), (1, 3), (2, 1), (2, 2), (2, 3)\}$;

   $B \times A = \{(1, 1), (2, 1), (3, 1), (1, 2), (2, 2), (3, 2)\}$;

   $(A \times B) \cap (B \times A) = \{(1, 1), (1, 2), (2, 1), (2, 2)\}$.

2 Denote $(A \times B) \cap (B \times A)$ by L, $(A \cap B) \times (A \cap B)$ by M. Note that if L and M are non-empty then a typical element of L or M is an *ordered pair*. Suppose L is non-empty; then

   $(x, y) \in L \Leftrightarrow (x, y) \in A \times B$ and $(x, y) \in B \times A$

   $\Leftrightarrow x \in A, y \in B$ and $x \in B, y \in A$

   $\Leftrightarrow x \in (A$ and $B), y \in (B$ and $A)$

   $\Leftrightarrow x \in A \cap B, y \in A \cap B$

   $\Leftrightarrow (x, y) \in (A \cap B) \times (A \cap B) = M$. Hence $L = M$.

3 For all $a \in A$ and $b \in B$, $(a, b) \in A \times B$. But $A \times B = C \times D$. Hence $(a, b) \in C \times D$, so that $a \in C$ and $b \in D$.Thus $(a \in A \Rightarrow a \in C)$ and $(b \in B \Rightarrow b \in D)$, so that $A \subseteq C$ and $B \subseteq D$. Similarly, for all $c \in C$ and $d \in D$, we can show that $c \in A$ and $d \in B$. Hence $C \subseteq A$ and $D \subseteq B$. Thus $A = C$ and $B = D$.

To show that it is not necessarily true that if $A \times B = C \times D = \varnothing$, then $A = C$ and $B = D$, consider the following counter-example:

Let $A = D = \varnothing$, $B = \{1\}$, $C = \{2\}$. Then $A \neq C$, $B \neq D$, but $A \times B = C \times D = \varnothing$.

4 $|A| = m, |B| = n \Rightarrow |A \times B| = mn.$

Proof $A \times B = \{(a, b): a \in A, b \in B\}.$
We have m choices for a. But b is chosen quite independently of a.
Hence there are n choices for b corresponding to each choice for a, giving
mn distinct pairs (a, b) in $A \times B$. Hence $|A \times B| = mn.$

5 See Fig. 1.10.

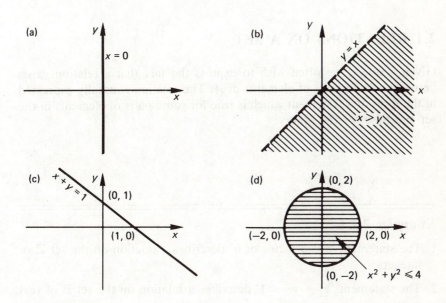

Fig. 1.10

6 (a) and (b) $\{(1, 4), (2, 4)\}$;
   (c) and (d) $\{(1, 3), (1, 4), (1, 5), (2, 3), (2, 4), (2, 5)\}.$

7 (a) Suppose $A \times (B \cap C)$ is non-empty. Then
   $(x, y) \in A \times (B \cap C) \Leftrightarrow x \in A, y \in B \cap C \Leftrightarrow x \in A,$
   $y \in (B \text{ and } C)$
   $\Leftrightarrow x \in A, y \in B \text{ and } x \in A, y \in C$
   $\Leftrightarrow (x, y) \in A \times B \text{ and } (x, y) \in A \times C$
   $\Leftrightarrow (x, y) \in (A \times B) \cap (A \times C).$
   (b) Suppose $A \times (B \cup C)$ is non-empty. $(x, y) \in A \times (B \cup C) \Leftrightarrow$
   $x \in A, y \in B \cup C \Leftrightarrow x \in A, y \in B \text{ or } y \in C \Leftrightarrow x \in A, y \in B$
   $\text{or } x \in A, y \in C \Leftrightarrow (x, y) \in A \times B \text{ or } (x, y) \in A \times C$
   $\Leftrightarrow (x, y) \in (A \times B) \cup (A \times C).$

2 RELATIONS

2.1 RELATIONS ON A SET

Given a set A, we often wish to express the fact that a relation exists between certain pairs of elements of A. The relation is usually expressed in the form of a statement which is true for some pairs of elements in the set and false for the others.

Examples 2.1.1

1. The statement 'a is a factor of b' describes a relation on the set $\mathbb{Z}$ of integers.

2. The statement '$x^2 + y^2 \leq 1$' describes a relation on the set $\mathbb{R}$ of real numbers.

3. The statement 'u is the uncle of v' describes a relation on the set P of all living people.

Notation

We denote a relation on a set A by the symbol $\boldsymbol{R}$ (the symbol $\sim$ is used by some authors, but we reserve this to denote an *equivalence* relation which is introduced in section 2.2). Let a, $b \in A$. Then we shall write $a \, \boldsymbol{R} \, b$ if 'a is related to b' is a *true* statement and $a \, \boldsymbol{\not R} \, b$ if 'a is related to b' is *false*.

Example 2.1.2

In example 2.1.1, 1, $A = \mathbb{Z}$ and R stands for 'is a factor of'.
Thus we write a R b to mean 'a is a factor of b'.
The following are examples of true statements:

   2 R 18;   3 R 3;   7 $\not{R}$ 90;   1 R m,   for every $m \in \mathbb{Z}$.

Properties of relations

Some relations have properties of special interest in mathematics and we
define three such properties in this section. First, however, we give in
example 2.1.3 a list of four different relations, all defined on the same
set for the sake of simplicity, and we use them to illustrate these properties.

Example 2.1.3

The following relations are all defined on the set $\mathbb{Z}$ of integers:

   (a) a R_1 b means $a < b$;
   (b) a R_2 b means $a + b \geq 5$;
   (c) a R_3 b means $a = 2b$;
   (d) a R_4 b means 'a and b have the same *parity*' (i.e. they are both
       even or both odd).

Reflexive property

A relation R on a set A is said to be *reflexive* if every element of A is
related to itself. That is R is reflexive $\Leftrightarrow$ a R a for every $a \in A$.

Example 2.1.4

We examine each of the relations on $\mathbb{Z}$ given above in examples 2.1.3 to
decide whether or not it is *reflexive*.

   (a) $a < a$ is not true for any $a \in \mathbb{Z}$. Hence R_1 is *not* reflexive.
   (b) $a + a \geq 5$ is not true for every $a \in \mathbb{Z}$; it is not true when $a = 2$,
       for example. Hence R_2 is *not* reflexive.

(c) $a = 2a$ is not true for every $a \in \mathbb{Z}$; it is not true when $a = 1$, for example. Hence $\boldsymbol{R}_3$ is *not* reflexive.

(d) Every integer a has the same parity as itself. Hence $\boldsymbol{R}_4$ is reflexive.

Notation

The symbol $\forall$ denotes 'for every' or 'for all'.

A statement about *every* member of a set (or every pair or triple of members of a set, etc.) is called a *universal statement* and the symbol $\forall$ is called the *universal quantifier*.

Example 2.1.5

The statements

(a) 1 is a factor of m, $\forall\ m \in \mathbb{Z}$;

(b) $a + a \geq 5$, $\forall\ a \in \mathbb{Z}$;

(c) $xy = yx$, $\forall\ x, y \in \boldsymbol{R}$;

(d) $x + y = 10 \Rightarrow y + x = 10$, $\forall\ x, y \in \boldsymbol{R}$;

are examples of *universal statements*; (a), (c) and (d) are *true* and (b) is *false*, since we can find an integer a for which the condition $a + a \geq 5$ does not hold.

Statement (c) reads '$xy = yx$, for *every* pair of real numbers x and y'; whereas statement (d) reads 'every pair of real numbers satisfying the condition $x + y = 10$ also satisfies the condition $y + x = 10$'.

Note carefully: To prove that a *universal statement* is

(a) TRUE: it is necessary to give a *general* argument which holds for *all* members (or pairs of members, etc.) of the set; it is *not* sufficient to verify that the statement is true for a few selected members of the set.

(b) FALSE: it is only necessary to find *one* member (or pair of members, etc.) of the set for which the statement is *false*; this is called 'giving a *counter-example*'.

We are interested in universal statements at this juncture because the statement 'the relation $\boldsymbol{R}$ on the set A is reflexive' is equivalent to the universal statement '$a\,\boldsymbol{R}\,a$, $\forall\ a \in A$'. The other two properties of relations that we now define are equivalent to universal statements of the type of example 2.1.5(d).

Symmetric property

A relation R on a set A is said to be *symmetric* if

$$a\,R\,b \Rightarrow b\,R\,a, \qquad \forall\ a, b \in A.$$

Example 2.1.6

We return to the relations given in example 2.1.3 above to decide whether or not each is *symmetric*.

(a) $a < b \not\Rightarrow b < a,\ \forall\ a, b \in \mathbb{Z}$.
   Counter-example: When $a = 2$ and $b = 3$, $a < b$ is true but $b < a$ is false. Hence R_1 is *not* symmetric.
(b) $a + b \geq 5 \Rightarrow b + a \geq 5,\ \forall\ a, b \in \mathbb{Z}$ and hence R_2 is symmetric.
(c) $a = 2b \not\Rightarrow b = 2a,\ \forall\ a, b \in \mathbb{Z}$.
   Counter-example: When $a = 2$, $b = 1$, $a = 2b$ is true, but $b = 2a$ is false.
   Hence R_3 is *not* symmetric.
(d) a has the same parity as $b \Rightarrow b$ has the same parity as $a,\ \forall\ a, b \in \mathbb{Z}$.
   Hence R_4 is symmetric.

Transitive property

A relation on a set A is said to be *transitive* if

$$a\,R\,b \quad \text{and} \quad b\,R\,c \Rightarrow a\,R\,c, \qquad \forall\ a, b, c \in A.$$

Example 2.1.7

We look again at each of the relations given in example 2.1.3 to decide whether or not it is *transitive*.

(a) $a < b$ and $b < c \Rightarrow a < c,\ \forall\ a, b, c \in \mathbb{Z}$. Hence R_1 is transitive.
(b) $a + b \geq 5$ and $b + c \geq 5 \not\Rightarrow a + c \geq 5,\ \forall\ a, b, c \in \mathbb{Z}$.
   Counter-example: when $a = 1$, $b = 6$, $c = 2$, $a + b \geq 5$ and $b + c \geq 5$ are true statements, but $a + c \geq 5$ is false.
   Hence R_2 is *not* transitive.

(c) $a = 2b$ and $b = 2c \not\Rightarrow a = 2c$, $\forall \ a, b, c \in \mathbb{Z}$.
Counter-example: when $c = 1$, $b = 2$ and $a = 4$, $a = 2b$ and
$b = 2c$ are true, but $a = 2c$ is false. Hence $\boldsymbol{R}_3$ is *not* transitive.

(d) $a \ \boldsymbol{R} \ b \Rightarrow a$ and b have the same parity;
$b \ \boldsymbol{R} \ c \Rightarrow b$ and c have the same parity.
Hence $a \ \boldsymbol{R}_4 \ b$ and $b \ \boldsymbol{R}_4 \ c \Rightarrow a$ and c have the same parity
$\Rightarrow a \ \boldsymbol{R}_4 \ c$, $\forall \ a, b, c \in \mathbb{Z}$. Hence $\boldsymbol{R}_4$ is transitive.

PROBLEMS 2.1a

1 Decide whether each of the following relations on the set given are
reflexive (R); symmetric (S); transitive (T). If you decide a relation
does not have one of the properties, give a counter-example.

(a) On the set $\mathbb{Z}^+$; $a \ \boldsymbol{R} \ b$ means a is a multiple of b.

(b) On the set $\mathbb{Z}$; $a \ \boldsymbol{R} \ b$ means ab is a multiple of 5.

(c) On the set $\mathbb{R}$; $x \ \boldsymbol{R} \ y$ means $x \neq y$.

(d) On the set $\mathbb{Z}$; $a \ \boldsymbol{R} \ b$ means $a - b$ is divisible by 2.

(e) On the set $\mathbb{Z}^+$; $a \ \boldsymbol{R} \ b$ means a and b have a common factor greater
than 1.

(f) On the set $\mathbb{Z}^+$; $a \ \boldsymbol{R} \ b$ means a and b have at least one digit in
common in the same decimal position, i.e. $1\underline{3} \ \boldsymbol{R} \ 79\underline{3}$, $\underline{1}3 \ \boldsymbol{R} \ \underline{2}11$ but
$13 \ \boldsymbol{\not R} \ 31$.

(g) On the set of lines in the plane; $l \ \boldsymbol{R} \ m$ means l is parallel to m.

(h) On the set of lines in the plane; $l \ \boldsymbol{R} \ m$ means l is perpendicular to
m.

(i) On the set of triangles in the plane; $\triangle_1 \ \boldsymbol{R} \ \triangle_2$ means $\triangle_1$ is
congruent to $\triangle_2$.

(j) On the set of points in the plane; let C be a fixed circle in the
plane. Define $X \ \boldsymbol{R} \ Y$ to mean the points X and Y lie either both
inside or both outside or both on the boundary of C.

2 (a) Give examples of two relations $\boldsymbol{R}_1$ and $\boldsymbol{R}_2$ on the set $\mathbb{Z}$ such that
$\boldsymbol{R}_1$ is reflexive and symmetric but not transitive and $\boldsymbol{R}_2$ is reflexive
and transitive but not symmetric.

(b) The relation $\boldsymbol{R}_3$ is defined on $\mathbb{Z}$ by $x \ \boldsymbol{R}_3 \ y$ means $x = y = 0$. Show
that $\boldsymbol{R}_3$ is symmetric and transitive but not reflexive.

Note: Taken together, these three examples $\boldsymbol{R}_1$, $\boldsymbol{R}_2$ and $\boldsymbol{R}_3$ show
that the reflexive, symmetric and transitive properties are *indepen-
dent*, since it is possible for a relation to possess any two without
the third.

(c) By considering $\boldsymbol{R}_3$ in 2(b) above, find the flaw in the following
argument: Let $\boldsymbol{R}$ be a symmetric and transitive relation on a set S.

Then if $x \in S$, $x \, \boldsymbol{R} \, y \Rightarrow y \, \boldsymbol{R} \, x$, since $\boldsymbol{R}$ is symmetric. But $x \, \boldsymbol{R} \, y \Rightarrow x \, \boldsymbol{R} \, x$, since $\boldsymbol{R}$ is transitive. Hence $\boldsymbol{R}$ is reflexive and we conclude symmetric + transitive ⇒ reflexive.

PROBLEMS 2.1b

1 Decide whether each of the following relations on the set given are reflexive (R), symmetric (S), transitive (T). Justify your answers by giving a brief proof or a counter-example, as appropriate.
   (a) On the set $\mathbb{R}$: $x \, \boldsymbol{R} \, y$ means $x - y$ is an integer.
   (b) On the set $\mathbb{R}$: $x \, \boldsymbol{R} \, y$ means $x + y$ is an integer.
   (c) On the set $\mathbb{Z}^+$: $a \, \boldsymbol{R} \, b$ means a and b have no common factor greater than 1.
   (d) On the set $\mathbb{Z}$: $a \, \boldsymbol{R} \, b$ means $a \geq |b|$.
   (e) On the set $\mathbb{Z} \times \mathbb{Z}$: $(x, y) \, \boldsymbol{R} \, (x_1, y_1)$ means $x + y_1 = y + x_1$.

2 A relation $\boldsymbol{R}$ on a set S is called *anti-symmetric* if $x \, \boldsymbol{R} \, y$ and $y \, \boldsymbol{R} \, x \Rightarrow x = y$, $\forall \, x, y \in A$.
   (a) Show that the relation on the set $\mathbb{Z}$ defined by $a \, \boldsymbol{R} \, b$ means 'a is a factor of b' is anti-symmetric.
   (b) Is the relation on $\mathbb{Z}$ defined by $a \, \boldsymbol{R} \, b$ means $a > b$ anti-symmetric?
   (c) Give an example of a relation on $\mathbb{Z}$ which is neither symmetric nor anti-symmetric.

2.2 EQUIVALENCE RELATIONS AND EQUIVALENCE CLASSES

Some of the relations we have met in examples 2.1.3 and problems 2.1 possess all three of the properties we have discussed, that is they are

(i) reflexive,   (ii) symmetric   and   (iii) transitive.

Such a relation is called an *equivalence relation* on the given set. We shall use the symbol ~ to denote an equivalence relation.

Three of the simplest examples of equivalence relations from section 2.1 are the relation of *congruence* on a set of triangles, *parallelism* on a set of lines and *having the same parity* on any subset of $\mathbb{Z}$. It is easy to verify that equality is also an equivalence relation on any set of numbers. In fact, equivalence can be regarded as a generalisation of the idea of equality. If we are talking about triangles and are interested only in their size and shape (and not in their positions, for instance) then any pair of congruent triangles are 'equivalent' in the sense of having 'the same' size and shape. Similarly, if we are discussing lines and are interested only in their direction, then any pair of parallel lines are 'equivalent' in the sense of having 'the same' direction.

PROBLEMS 2.2.1b

1 Show that *similarity* is an equivalence relation on the set of all triangles in the plane. Suggest a further equivalence relation on this set (i.e. other than congruence and similarity).

2 Suggest at least two equivalence relations on the set of all convex plane polygons.

3 Let $\sim$ be a relation on $\mathbb{Z}^+$ defined by $m \sim n$ means that m and n have the *same number of digits*. Prove that $\sim$ is an equivalence relation on $\mathbb{Z}^+$. Suggest some other equivalence relations on $\mathbb{Z}^+$ or $\mathbb{Z}$.

Equivalence classes

The idea of equivalence leads very naturally to the idea of grouping together all those elements which are equivalent (i.e. related by the equivalence relation) to one another. For example, consider the equivalence relation of parallelism on a set S of lines in a plane. If we pick any line l in S we can class together with l all the lines parallel to l to form a subset $L \subseteq S$.

Every line in L is parallel (and hence related) to every other line in L, and is not parallel (and hence not related) to any line not in L. We call this set the *equivalence class* of l and denote it by $\bar{l}$. The line l is called a *class representative* of this equivalence class. Now within the set $L = \bar{l}$, there is nothing special about l. If m is any other line in L, then every line of L is also parallel to m and all the lines in S which are parallel to m are in L. Hence $L = \bar{m}$, and m could equally well be chosen as a class representative for the *same* equivalence class. This property of equivalence classes, that any pair of related elements determine the *same* equivalence class, is true for *any* equivalence relation on *any* set; we shall prove this later in this section. For the moment, we take this as intuitive and look at the problem of determining the equivalence classes in some practical examples.

Examples 2.2.1

1. The relation $\sim$ defined on the set $\mathbb{Z}^+$ by $m \sim n$ means that m and n have the same number of digits is clearly an equivalence relation (check!). We will determine the equivalence classes.

$\mathbb{Z}^+$ lends itself to being treated systematically, so we start by finding $\bar{1}$. Now 1 has just one digit and hence

$$E_1 = \bar{1} = \{1, 2, 3, \ldots, 9\} = \{n \in \mathbb{Z}: 1 \leq n \leq 9\}.$$

As in the discussion above, we see that $\bar{2} = \bar{3} = \ldots = \bar{9} = E_1$ and hence the least number left to deal with is 10. Since 10 has two digits,

$$E_2 = \overline{10} = \{10, 11, 12, \ldots, 99\} = \{n \in \mathbb{Z}: 10 \leq n \leq 99\}.$$

Similarly,

$$E_3 = \{100, 101, 102, \ldots, 999\} = \{n \in \mathbb{Z}: 100 \leq n \leq 999\}.$$

Hence the equivalence classes are E_1, E_2, ..., E_r, ..., where $E_r = \{n \in \mathbb{Z}: 10^{r-1} \leq n \leq 10^r - 1\}$.

2. The relation $\sim$ defined on $\mathbb{R}$ by $x \sim y$ means that $x - y \in \mathbb{Z}$ is an equivalence relation (check!). We determine the equivalence classes.

This time we cannot be so systematic. To get an idea of how the equivalence classes are formed we will take any real number x and find $\bar{x}$. We then hope to be able to generalise to give the form of all the equivalence classes, as in the previous example.

Suppose we take $x = 1.73$. Then $y \in \bar{x}$ if $y \sim x$, i.e. if $y - x$ is an integer.

$y - x$	...	-3	-2	-1	0	1	2	...
x	...	-1.27	-0.27	0.73	1.73	2.73	3.73	...

Hence $\bar{x} = \{\ldots, -1.27, -0.27, 0.73, 1.73, 2.73, \ldots\}$.
Generalising, let $\alpha \in \mathbb{R}$ be any real number; then

$$\bar{\alpha} = \{\ldots, \alpha - 2, \alpha - 1, \alpha, \alpha + 1, \alpha + 2, \ldots\}$$
$$= \{\alpha + r: r \in \mathbb{Z}\}.$$

PROBLEMS 2.2a

1 Check that each of the following relations $\sim$ are equivalence relations on the given set. Describe in words a typical equivalence class.

For each example complete the statement: 'There is an equivalence class corresponding to each ....'

(a) The set of residents of a given town; $a \sim b$ means that a and b live in the same street.

(b) The same set as in (a); $a \sim b$ means a and b were born in the same year.

(c) The same set as in (a); $a \sim b$ have their birthday on the same day of the year.

(d) The set of animals in London Zoo; $x \sim y$ means x and y belong to the same species.

(e) The set of paintings in the National Gallery; $p \sim q$ means that p and q were painted by the same artist.

(f) The set of students enrolled on any degree course at a given university: $s \sim t$ means s and t are enrolled for the same degree course.

2 (a) Find the equivalence classes determined by the equivalence relation $\sim$ on the set $\mathbb{Z}$ defined by $x \sim y$ means x and y have the same parity (see section 2.1).

(b) $\sim$ is the relation defined on $\mathbb{Z}$ by $x \sim y$ means that $|x| = |y|$. Check that $\sim$ is an equivalence relation on $\mathbb{Z}$ and find the equivalence classes.

3 The relation defined on $\mathbb{Z} \times \mathbb{Z}$ by $(x, y) \sim (x', y') \Leftrightarrow x + y' = y + x'$ is an equivalence relation (see problem 2.1b, 1(e)).

Find (a) $\overline{(1, 2)}$; (b) $\overline{(0, -3))}$; (c) $\overline{(a, b)}$; describe this set geometrically on a cartesian diagram.

4 The relation defined on $\mathbb{Z}^+$ by $m \sim n \Leftrightarrow mn$ is the square of an integer is an equivalence relation.

Find (a) $\overline{1}$; (b) $\overline{2}$; (c) $\overline{p}$, where p is a prime number.

5 Let $A = \{2, 3, 4, 6\}$. A relation $\boldsymbol{R}$ is defined on the set A by $x \boldsymbol{R} y \Leftrightarrow x < y$. Corresponding to each element $a \in A$, we define a subset $\boldsymbol{R}(a)$ by

$$\boldsymbol{R}(a) = \{x \in A : x \boldsymbol{R} a\}.$$

(a) Find the subsets $\boldsymbol{R}(a)$ for each $a \in A$.

(b) Give an example of an element $a \in A$ such that $\boldsymbol{R}(a) = \varnothing$.

(c) Give an example of a pair of elements $a, b \in A$ such that $\boldsymbol{R}(a) \cap \boldsymbol{R}(b) \neq \varnothing$.

(d) Show that $\boldsymbol{R}$ is *not* an equivalence relation. Do you think it would be possible to find examples satisfying (b) and (c) for an *equivalence* relation $\sim$ on A?

PROBLEMS 2.2b

1 Let $\sim$ be the relation on $\mathbb{Z}^+$ defined by $m \sim n$ means that m and n terminate in the same digit. Prove that $\sim$ is an equivalence relation on $\mathbb{Z}^+$ and find the equivalence classes.

2 A relation $\sim$ is defined on $\mathbb{Z}^+$ by $a \sim b \Leftrightarrow$ there exists $k \in \mathbb{Z}$ such that $a = 10^k b$.

Prove that $\sim$ is an equivalence relation.

Find (a) $\bar{1}$;   (b) $\overline{200}$;   (c) $\overline{170}$.

3 Let $\sim$ be the relation on $\mathbb{R}$ defined by $x \sim y \Leftrightarrow [x] = [y]$, where the symbol $[x]$ is used to denote the greatest integer less than or equal to x (i.e. $[0.13] = 0$, $[-8.7] = -9$, $[2] = 2$).

Prove that $\sim$ is an equivalence relation on $\mathbb{R}$ and define the equivalence classes.

4 An equivalence relation $\sim$ on the set $\mathbb{C}$ of complex numbers is defined by $w \sim z$ means w and z have the same modulus. Describe the equivalence class containing (a) $2i$; (b) $1 + i$. Indicate these equivalence classes on a sketch of the Argand diagram.

5 Let $\mathbb{Z}^*$ denote the set of non-zero integers. A relation $\sim$ is defined on $\mathbb{Z} \times \mathbb{Z}^*$ by

$$(a_1, b_1) \sim (a_2, b_2) \Leftrightarrow a_1 b_2 = a_2 b_1.$$

(a) Prove that $\sim$ is an equivalence relation on $\mathbb{Z} \times \mathbb{Z}^*$.
(b) Show that for any non-zero integer k, $(a, b) \sim (ak, bk)$, $\forall$ (a, b) $\in \mathbb{Z} \times \mathbb{Z}^*$.
(c) Find the following equivalence classes:

$$\overline{(2, 3)}; \quad \overline{(1, 1)}; \quad \overline{(0, c)}; \quad \overline{(a, b)}, \quad a \neq 0.$$

6 Let $A = \{r \in \mathbb{Z}: 1 \leq r \leq 9\}$. A relation R is defined on A by $a \, R \, b \Leftrightarrow a + b \leq 8$. Corresponding to each element $a \in A$, we define a subset $R(a)$ of A by

$$R(a) = \{x \in A, x \, R \, a\}.$$

(a) Find the subsets $R(a)$ for each $a \in A$.
(b) Find an element $a \in A$ for which $R(a) = \emptyset$.
(c) Give an example of a pair of elements $a, b \in A$ for which $R(a) \cap R(b) \neq \emptyset$.
(d) Prove that R is *not* an equivalence relation.

Properties of equivalence classes

Earlier in this section, we took it as intuitive in the examples of equivalence relations that any pair of related elements determine the *same* equivalence class. We shall now show how to use the symmetric and transitive properties of an equivalence relation to prove that this is valid for any equivalence relation on the set. First, a formal definition of an equivalence class.

Equivalence class: formal definition

Let $\sim$ be an equivalence relation on a set S. We define the *equivalence class* $\bar{a}$ of an element $a \in S$ to be the subset $\bar{a} = \{x \in S: x \sim a\}$.

RESULT 2.2.1   Let $\sim$ be an equivalence relation on a set S. Then
$$a \sim b \Rightarrow \bar{a} = \bar{b}, \qquad \forall\, a, b \in S.$$

Proof   To show $\bar{a} = \bar{b}$, we must prove that (i) $x \in \bar{a} \Rightarrow x \in \bar{b}$ and (ii) $y \in \bar{b} \Rightarrow y \in \bar{a}$ (see section 1.1 on equality of sets).

Now $\bar{a} = \{x \in S: x \sim a\}$; $\bar{b} = \{y \in S: y \sim b\}$.

(i) $x \in \bar{a} \Rightarrow x \sim a$, by definition of $\bar{a}$.
  Also we have $a \sim b$, given.
  But $x \sim a$ and $a \sim b \Rightarrow y \sim b$, by the *transitive* property of $\sim$,

$$\Rightarrow x \in \bar{b} \Rightarrow y \sim b, \text{ by definition of } \bar{b}.$$

(ii) $y \in \bar{b} \Rightarrow y \sim b$, by definition of $\bar{b}$.
  Also $a \sim b \Rightarrow b \sim a$, by the *symmetric* property of $\sim$.
  But $y \sim b$ and $b \sim a \Rightarrow y \sim a$, by the *transitive* property of $\sim$,

$$\Rightarrow y \in \bar{a}, \text{ by definition of } \bar{a}.$$

Hence from (i) and (ii), $\bar{a} = \bar{b}$. ∎

It is easy to show that the converse of this result also holds.

RESULT 2.2.2 $\bar{a} = \bar{b} \Rightarrow a \sim b, \forall\, a, b \in S.$

Proof   Suppose $\bar{a} = \bar{b}$.
We know $a \in \bar{a}$, by the *reflexive* property of $\sim$.
Hence $a \in \bar{b}$, since $\bar{a} = \bar{b}$.
But $a \in \bar{b} \Rightarrow a \sim b$, by definition of $\bar{b}$. ∎

Combining results 2.2.1 and 2.2.2 gives that for any equivalence relation $\sim$ on any set S,

$$a \sim b \Leftrightarrow \bar{a} = \bar{b}.$$

This result justifies our earlier assumption that any element of an equivalence class can equally well serve as class representative. Note that we have had to use all three properties of an equivalence relation in the proof.

A further property of equivalence classes that you will probably have noticed is that any pair of distinct equivalence classes are disjoint. Our original example of parallelism as an equivalence relation on a set of lines S makes this clear. Let $l \in S$ be any line. Then the equivalence class $\bar{l}$ of l is the set consisting of all the lines of S parallel to l. Thus if $\bar{m}$ is an equivalence class distinct from $\bar{l}$, then $\bar{m}$ must consist of all the lines parallel to some other direction. Clearly, no line can be parallel to two different directions at once and hence $\bar{l} \cap \bar{m} = \varnothing$.

We now prove this property holds for any equivalence relation on any set S.

RESULT 2.2.3 Let $\sim$ be an equivalence relation on a set S. Let $\bar{a}$ and $\bar{b}$ be two distinct equivalence classes.

Then $$\bar{a} \cap \bar{b} = \varnothing.$$

Proof We use the method of *contradiction* or *reductio ad absurdum* for this proof; that is, we ask what happens if the result is false and we can find an element common to $\bar{a}$ and $\bar{b}$? We then show that this leads to a contradiction of the original information we are given (in this case, that $\bar{a} \neq \bar{b}$).

Suppose if possible that there exists an element $w \in \bar{a} \cap \bar{b}$.
Then $w \in \bar{a} \Rightarrow w \sim a \Rightarrow \bar{w} = \bar{a}$, by result 2.2.1.
Similarly, $w \in \bar{b} \Rightarrow w \sim b \Rightarrow \bar{w} = \bar{b}$.
Hence $w \in \bar{a} \cap \bar{b} \Rightarrow \bar{a} = \bar{b} (= \bar{w})$.
But this contradicts the condition $\bar{a} \neq \bar{b}$. Hence no such element $w \in \bar{a} \cap \bar{b}$ can exist. Thus $\bar{a} \neq \bar{b} \Rightarrow \bar{a} \cap \bar{b} = \varnothing$. ∎

To sum up: given an equivalence relation $\sim$ on a set S and any pair of elements $a, b \in S$, then

(a) either $\bar{a} = \bar{b}$ or $\bar{a} \cap \bar{b} = \varnothing$ (from result 2.2.3);
(b) $a \sim b \Leftrightarrow \bar{a} = \bar{b}$ from results 2.2.1 and 2.2.2);
     and hence also
(c) $a \not\sim b \Leftrightarrow \bar{a} \neq \bar{b} \Leftrightarrow \bar{a} \cap \bar{b} = \varnothing$.

2.3 PARTITIONS AND EQUIVALENCE CLASSES

In section 1.3 we defined a *partition* of a set S as a collection Σ of *cells* (non-empty subsets of S) such that every element of S was a member of one and only one of the cells in Σ. In this section we shall show that a collection of equivalence classes partitions a set and conversely every partition can be viewed as defining an equivalence relation.

The following examples of partitions arise from equivalence relations defined in the examples or problems in the preceding section.

Examples 2.3.1

1. Let $X_1 = \{1, 3, 5, 7, \ldots\} = \{2r - 1: r \in \mathbb{Z}^+\}$
   and $X_2 = \{2, 4\ 6, 8, \ldots\} = \{2r: r \in \mathbb{Z}^+\}$.
   Then $\Sigma = \{X_1, X_2\}$ is a partition of $\mathbb{Z}^+$.

2. Let $X_1 = \{1, 11, 21, \ldots\}$, $X_2 = \{2, 12, 22, \ldots\}$, $\ldots$,
   $X_r = \{r, 10 + r, \ldots\}$, $\ldots$, $X_{10} = \{10, 20, 30, \ldots\}$.
   Then $\Sigma = \{X_r: 1 \leq r \leq 10\}$ is a partition of $\mathbb{Z}^+$.

3. Let $X_1 = \{1, 2, \ldots, 9\}$, $X_2 = \{10, 11, \ldots, 99\}$,
   $X_3 = \{100, 101, \ldots, 999\}$, $\ldots$, $X_r = \{n: 10^{r-1} \leq n < 10^r\}$, $\ldots$.
   Then $\Sigma = \{X_r: r \in \mathbb{Z}^+\}$ is a partition of $\mathbb{Z}^+$.

THEOREM 2.3.1   Let $\sim$ be an equivalence relation on a set S and let $\bar{x}$ denote the equivalence class of x. Then the collection $\Sigma = \{\bar{x}: x \in S\}$ is a partition of S.

Proof   Now $x \sim x$, $\forall\ x \in S$, since $\sim$ is reflexive.
Hence $x \in \bar{x}$, $\forall\ x \in S$.
Thus (i) all the members of Σ are non-empty and (ii) every element of S belongs to a member of Σ.
From result 2.2.3, we know that $\bar{x} \neq \bar{y} \Rightarrow \bar{x} \cap \bar{y} = \emptyset$.
Hence (iii) every pair of distinct members of Σ are disjoint. The statements (i), (ii), and (iii) establish that Σ is a partition of S. ∎

The following theorem is the converse of theorem 2.3.1. It explains how a partition of a set S can be used to define an equivalence relation on S, such that the equivalence classes are the cells forming the partition. The proof is simple and is left as an exercise.

THEOREM 2.3.2   Let S be a set and Σ a partition of S. Define a relation on S by the rule $x \sim y \Rightarrow x$ and y belong to the same cell in Σ. Then $\sim$ is an equivalence relation on S and the equivalence classes are just the cells.

PROBLEMS 2.3a

1 For the examples 1 and 2 in examples 2.3.1 above, sketch a Venn diagram illustrating how Σ partitions $\mathbb{Z}^+$.

2 For the examples 1 to 3 in examples 2.3.1 above, define an equivalence relation $\sim$ on $\mathbb{Z}^+$ such that the equivalence classes are the cells in Σ.

3 Let $k \in \mathbb{Z}$.
We define a subset E_k of $\mathbb{R}$ by $E_k = \{x \in \mathbb{R}: k \le x < k + 1\}$. Show that $\Sigma = \{E_k: k \in \mathbb{Z}\}$ is a partition of $\mathbb{R}$. Define an equivalence relation $\sim$ on $\mathbb{R}$ such that the equivalence classes are the cells in Σ.

4 Define an equivalence relation $\sim$ on the set $A = \{a, b, c, d\}$ such that the equivalence classes are $\{a, b, c\}$ and $\{d\}$.

PROBLEMS 2.3b

1 Prove theorem 2.3.2.

2 Let S be a set and $x \in S$. The subset E_x of S is defined by $E_x = \{x\}$. Show that $\Sigma = \{E_x: x \in S\}$ is a partition of S. What is the equivalence relation on S that has the cells in Σ as equivalence classes?

3 Find the number of different partitions of a set with three elements.

2.4 CONGRUENCE RELATIONS ON THE INTEGERS

In this section we introduce a particular type of equivalence relation on the set $\mathbb{Z}$ of integers. We shall first give a formal definition of *divisibility* of integers.

Divisors and multiples

Let $a, d \in \mathbb{Z}$. We say that a is divisible by d (or that a is a *multiple* of d) if there is an integer $k \in \mathbb{Z}$ such that $a = kd$. Then d is called a *divisor* of a.

Examples 2.4.1

1. Let S be the set of integers divisible by 4. Then

$$S = \{4r: r \in \mathbb{Z}\}.$$

2. Let P be the set of multiples of an integer $m \in \mathbb{Z}$. Then

$$P = \{mk: k \in \mathbb{Z}\}$$

Congruence of integers

Let m be some fixed positive integer. Let $a, b \in \mathbb{Z}$; then we say that 'a is congruent to $b (modulo \ m)$' if and only if $a - b$ is divisible by m.

Notation

The statement 'a is congruent to $b (modulo \ m)$' is written symbolically as

$$a \equiv b \pmod{m}.$$

Equivalent statements of the definition of congruence

We give below three different statements of the definition of congruence. You should convince yourself that we are really saying the same thing in three slightly different ways; that is, any pair of integers which are congruent by any one definition will also be congruent by either of the other two.

(a) $a \equiv b \pmod{m} \Leftrightarrow$ there exists $k \in \mathbb{Z}$ such that $a - b = km$.
(This is just the definition given above written symbolically.)

(b) $a \equiv b \pmod{m} \Leftrightarrow a$ and b leave the same remainder on division by m.

(c) $a \equiv b \pmod{m} \Leftrightarrow$ there exists $k \in \mathbb{Z}$ such that $a = b + km$.
(This is just a slight rearrangement of definition (a).)

Examples 2.4.2

1. $33 \equiv 23 \equiv 193 \equiv 3 \pmod{10}$.

2. $2 \equiv -5 \equiv -12 \equiv 23 \pmod{7}$.

Congruence relations have proved an important and fruitful tool in algebra as well as in number theory. In Chapter 4 we look at the arithmetic of congruence relations in detail. At the moment our interest in them is explained by the following theorem.

THEOREM 2.4.1 Let m be a fixed positive integer. Then the relation $\sim$ defined on $\mathbb{Z}$ by $a \sim b \Leftrightarrow a \equiv b \,(\text{mod } m)$ is an equivalence relation.

Proof

(i) We show $\sim$ is *reflexive*. Let $a \in \mathbb{Z}$. Then $a - a = 0 = 0m$. Hence $a \equiv a \,(\text{mod } m)$ and so $a \sim a$, $\forall\, a \in \mathbb{Z}$.

(ii) We show $\sim$ is *symmetric*. Let $a, b \in \mathbb{Z}$ and suppose $a \sim b$. Then $a - b = km$ for some $k \in \mathbb{Z}$. But $a - b = km \Rightarrow b - a = (-k)m \Rightarrow b \equiv a \,(\text{mod } m)$, since $-k \in \mathbb{Z}$. Hence $a \sim b \Rightarrow b \sim a$, $\forall\, a, b \in \mathbb{Z}$.

(iii) We show $\sim$ is *transitive*. Let $a, b, c \in \mathbb{Z}$ and suppose $a \sim b$ and $b \sim c$. Then there exist integers $k, h \in \mathbb{Z}$ such that $a - b = km$ and $b - c = hm$. Then $a - c = (k + h)m$ and hence $a \equiv c \,(\text{mod } m)$, since $k + h \in \mathbb{Z}$. Thus $a \sim b$ and $b \sim c \Rightarrow a \sim c$, $\forall\, a, b, c \in \mathbb{Z}$.

Congruence classes

The equivalence classes determined by the relation of congruence mod m on $\mathbb{Z}$ are called the *congruence* (or *residue*) *classes* mod m.

Examples 2.4.3

1. We determine the congruence classes mod 4.

$\bar{0} = \{x \in \mathbb{Z}: x \equiv 0 \,(\text{mod } 4)\}$
$\quad = \{x \in \mathbb{Z}: x = 4k, k \in \mathbb{Z}\} = \{\ldots, -12, -8, -4, 0, 4, 8, 12, \ldots\}.$
$\bar{1} = \{x \in \mathbb{Z}: x = 1 + 4k, k \in \mathbb{Z}\} = \{\ldots, -11, -7, -3, 1, 5, 9, 13, \ldots\}.$
$\bar{2} = \{x \in \mathbb{Z}: x = 2 + 4k, k \in \mathbb{Z}\} = \{\ldots, -10, -6, -2, 2, 6, 10, \ldots\}.$
$\bar{3} = \{x \in \mathbb{Z}: x = 3 + 4k, k \in \mathbb{Z}\} = \{\ldots, -9, -5, -1, 3, 7, 11, 15, \ldots\}.$

It is easy to see that every integer belongs to one and only one of these four congruence classes; i.e. the congruence classes $\{\bar{0}, \bar{1}, \bar{2}, \bar{3}\}$ *partition* $\mathbb{Z}$.

47

2. Find the least integer $x \geq 0$ satisfying the congruence relation

$$x - 5 \equiv -1 (\text{mod } 3).$$

Solution
Method (1) $x - 5 \in \overline{-1} (\text{mod } 3)$
$$= \{\ldots, -7, -4, -1, 2, 5, 8, \ldots\}.$$
Hence $x \in \{\ldots, -7+5, -4+5, -1+5, 2+5, 5+5, \ldots\}$
$$= \{\ldots, -2, \textcircled{1}, 4, 7, 10, \ldots\}.$$

Thus the least integer $x \geq 0$ satisfying the congruence relation is $x = 1$.

Method (2) We could tackle this type of problem another way by turning it into an equation.

$$x - 5 \equiv -1 (\text{mod } 3) \Rightarrow x - 5 = -1 + 3k, \quad \text{for some } k \in \mathbb{Z}$$
$$\Rightarrow x = 4 + 3k, \quad \text{for some } k \in \mathbb{Z}.$$

In order to find the least $x \geq 0$, we need to choose $k = -1$, giving $x = 4 - 3 = 1$, as before.

3. Find the least integer $x \geq 0$ satisfying the congruence

$$2x \equiv -7 (\text{mod } 5).$$

Solution
Method (1) $2x \in \overline{-7} (\text{mod } 5) = \{\ldots, -12, -7, -2, 3, 8, 13, \ldots\}.$

Since x is an integer and $x \geq 0$, we look for the least positive integer in this set divisible by 2, which gives

$$2x = 8, \qquad x = 4.$$

Method (2) Using the equation approach, we could write:

$$2x \equiv -7 (\text{mod } 5) \Rightarrow 2x = -7 + 5k, \quad \text{for some } k \in \mathbb{Z}.$$

Since $x \geq 0$, we need only consider values of $k > 1$.
$k = 2$ gives $2x = 3$, which does not give an integer for x.
$k = 3$ gives $2x = 8 \Rightarrow x = 4$, which is thus the least positive integer solution.

PROBLEMS 2.4a

1 Find the congruence classes (a) mod 2 and (b) mod 5.

2 Let $a \in \mathbb{Z}$. Find the congruence class $\overline{a}$ of a (a) mod 3 and (b) mod m, where m is a fixed positive integer.

3 (a) Find the least integer $x \geq 0$ satisfying each of the following congruence relations.

(i) $x \equiv 17 \pmod 3$; (ii) $x \equiv -9 \pmod 5$; (iii) $x \equiv -6 \pmod 3$;
(iv) $x \equiv -7 \pmod 2$; (v) $x - 1 \equiv 57 \pmod{10}$.

(b) Find the set of all solutions to each of the congruence relations in (a).

4 Find the least positive integer x_0 such that $x = x_0$ satisfies $2x \equiv 5 \pmod 9$. Let $y \in \mathbb{Z}$ be any integer such that $y \equiv x_0 \pmod 9$. Prove that $2y \equiv 5 \pmod 9$.

5 Explain why there is no integer x such that $5x \equiv 1 \pmod{10}$.

6 Let m be a positive integer and $a, b \in \mathbb{Z}$ be such that $a \equiv b \pmod m$. Prove that

(a) $a + c \equiv b + c \pmod m$, $\forall c \in \mathbb{Z}$;
(b) $ac \equiv bc \pmod m$, $\forall c \in \mathbb{Z}$.

PROBLEMS 2.4b

1 Find the solution set for each of the following congruence relations and show that in each case the set is a congruence class.

(a) $x - 3 \equiv 13 \pmod 5$;   (b) $32 \equiv x \pmod 6$;
(c) $17 - x \equiv 1 \pmod 8$;   (d) $x + 53 \equiv 0 \pmod{19}$.

2 In each of the congruence relations $ax \equiv b \pmod m$ listed in (i) to (iv) below

(a) find the least integer $x_0 \geq 0$ satisfying the congruence;
(b) verify that any integer y such that $y \equiv x_0 \pmod m$ also satisfies $ay \equiv b \pmod m$;

(i) $4x \equiv 3 \pmod{17}$;   (ii) $3x + 1 \equiv 9 \pmod{11}$;
(iii) $6x - 1 \equiv 0 \pmod 7$:   (iv) $3 \equiv 2x \pmod 5$.

3 For each of the following congruence relations find, where possible, the least integer $x \geq 0$ satisfying the congruence

(a) $2x \equiv 3 \pmod 6$;   (b) $5x \equiv 3 \pmod 6$;   (c) $3x \equiv 2 \pmod 6$.

4 Let m be a positive integer and $a_1, a_2, b_1, b_2 \in \mathbb{Z}$ be such that $a_1 \equiv b \pmod m$ and $a_2 \equiv b_2 \pmod m$. Prove that

(a) $a_1 \pm a_2 \equiv b_1 \pm b_2 \pmod m$;   (b) $a_1 a_2 \equiv b_1 b_2 \pmod m$.

5 Let m be a positive integer and $a \in \mathbb{Z}$. Prove that $x \equiv a \pmod m$ $\Rightarrow$ $x^2 \equiv a^2 \pmod m$.

Using the fact that every integer belongs to one of the congruence classes $\bar{0}, \bar{1}, \bar{2}, \ldots, \bar{9} \pmod{10}$, prove that the square of every integer terminates in one of the digits 0, 1, 4, 5, 6 or 9.

Explain why it is impossible to find an integer n such that n^4 terminates in the digit 9.

Can anything be said about the final digit of the cube of an integer?

6 Using the method of problem 5 with $m=4$, show that given any integer n, either n^2 is divisible by 4, or $n^2 = 4k + 1$ for some $k \in \mathbb{Z}$.

2.5 THE GRAPH OF A RELATION

In this section we shall look at a way of illustrating a relation $\boldsymbol{R}$ on a set S.

Suppose we let a set of points represent the elements of S and we joint two points a, $b \in S$ by a *directed arc*, as shown in Fig. 2.1(a), if $a \boldsymbol{R} b$. If both $a \boldsymbol{R} b$ and $b \boldsymbol{R} a$, then a and b are joined by two arcs with opposite orientations as shown in Fig. 2.1(b). Finally, if $a \boldsymbol{R} a$, then a is joined to itself by a *loop* as shown in Fig. 2.1(c).

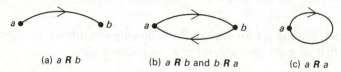

(a) $a \boldsymbol{R} b$ (b) $a \boldsymbol{R} b$ and $b \boldsymbol{R} a$ (c) $a \boldsymbol{R} a$

Fig. 2.1

Directed graph

A set S of points together with a set A of directed arcs (including loops) joining some pairs of the points is known as a *directed graph*.

We denote an arc directed from a point a to a point b by the ordered pair (a, b) and a loop joining a point c to itself by (c, c). Thus the set of directed arcs A can be thought of as a subset of the cartesian product set $S \times S$ (see section 1.5):

$$A = \{(a, b) \in S \times S : a \boldsymbol{R} b\}.$$

The discussion above shows that any relation $\boldsymbol{R}$ on a set S can be used to define a directed graph and conversely any directed graph can be used to define a relation on its set of points.

Examples 2.5.1

1. Let $S = \{1, 2, 3, 4\}$ and $\boldsymbol{R}_1$ be defined on S by

$$a_1 \boldsymbol{R}_1 b \Leftrightarrow a \leq b.$$

The set of arcs is given by

$A = \{(1, 1), (1, 2), (1, 3), (1, 4), (2, 2), (2, 3), (2, 4), (3, 3), (3, 4), (4, 4)\}.$

(See Fig. 2.2.)

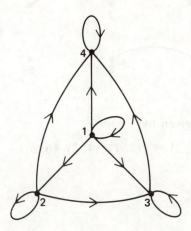

Fig. 2.2

2. Let $S = \{1, 2, 3, 4\}$ and R_2 be defined on S by

$$a\ R_2\ b \Leftrightarrow a + b \leq 5.$$

The set of arcs is given by

$A = \{(1, 1), (1, 2), (1, 3), (1, 4), (2, 1), (2, 2), (2, 3), (3, 1), (3, 2), (4, 1)\}.$

(See Fig. 2.3.)

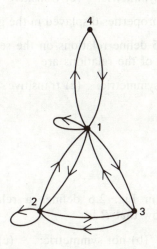

Fig. 2.3

3. Let $S = \{1, 2, 3, 4, 5, 6\}$ and $\boldsymbol{R_3}$ be defined on S by

$$a \; \boldsymbol{R_3} \; b \Leftrightarrow a = 2b.$$

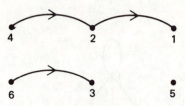

4   2   1

6   3   5

Fig. 2.4

The set of arcs is given by

$$A = \{(2, 1), (4, 2), (6, 3)\}.$$

(See Fig. 2.4.)

PROBLEMS 2.5a

1 Draw graphs to illustrate the following relations on the set
$S = \{1, 2, 3, 4, 5, 6\}$.

(a) $a \; \boldsymbol{R_1} \; b \Leftrightarrow a = b$;   (b) $a \; \boldsymbol{R_2} \; b \Leftrightarrow a = b - 1$;
(c) $a \; \boldsymbol{R_3} \; b \Leftrightarrow a \equiv b \,(\mathrm{mod}\ 2)$;   (d) $a \; \boldsymbol{R_4} \; b \Leftrightarrow a$ is a factor of b.

2 Which of the relations of problem 1 are

(a) reflexive;   (b) symmetric;   (c) transitive?

How is each of these properties displayed in the graph of a relation $\boldsymbol{R}$?

3 The graphs in Fig. 2.5 define relations on the set $S = \{a, b, c, d\}$.
Decide which, if any, of the relations are

(a) reflexive;   (b) symmetric;   (c) transitive.

PROBLEMS 2.5b

1 The directed graph in Fig. 2.6 defines a relation $\boldsymbol{R}$ on the sets
$S = \{w, x, y, z\}$. Show that $\boldsymbol{R}$ is

(a) not reflexive;   (b) not symmetric;   (c) not transitive.

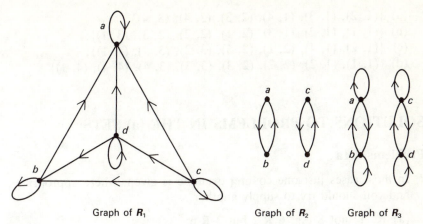

Graph of R_1 Graph of R_2 Graph of R_3

Fig. 2.5

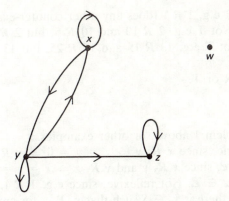

Fig. 2.6

2 Give the *minimal* set of arcs which need to be added to the graph illustrated in Fig. 2.6 in order that it defines an equivalence relation. What are then the equivalence classes?

3 State a condition on the ordered pairs in the arc set A of the graph of a relation R on a set S if R is to be

  (a) reflexive;   (b) symmetric;   (c) transitive.

4 The condition for a relation R on a set S to be called *anti-symmetric* is given in problem 2.1b, 2. Explain how this property is displayed in the graph of an anti-symmetric relation R.

    Draw the graph of each of the relations on the set $S = \{1, 2, 3, 4\}$ whose arc sets are given below. For each, decide whether it is symmetric, anti-symmetric or neither.

(a) {(1, 2), (1, 3), (1, 4), (2, 3), (2, 4), (3, 4)};
(b) {(1, 1), (1, 2), (1, 3), (2, 1), (2, 2), (2, 3), (3, 1)};
(c) {(1, 2), (1, 4), (2, 1), (2, 3), (3, 2), (3, 3), (4, 1)};
(d) {(1, 1), (1, 2), (2, 2), (2, 3), (3, 3), (3, 4), (4, 4), (4, 1)}.

SOLUTIONS TO PROBLEMS IN THE (a) SETS

Problems 2.1a

1 In most cases just one counter-example is given, where appropriate, and you should try to supply another.

(a) R, T; not S e.g. 6 R 2 but 2 $\not R$ 6.
(b) S; not R e.g. 1 $\not R$ 1. Not T e.g. 2 R 5 and 5 R 1 but 2 $\not R$ 1.
(c) S; not R e.g. any $x \in \mathbb{R}$. Not T e.g. 1 R 2 and 2 R 1, but 1 $\not R$ 1.
(d) R, S, T.
(e) S. Not R e.g. 1 $\not R$ 1 (does any other counter-example exist in this case?). Not T e.g. 2 R 10 and 10 R 5, but 2 $\not R$ 5.
(f) R, S. Not T e.g. 13 R 15 and 15 R 25, but 13 $\not R$ 25.
(g) R, S, T.
(h) S. Not R or T.
(i) R, S, T.
(j) R, S, T.

2 (a) See problem 1 above, or other examples.
(b) Symmetric, since $x\ R_3\ y \Rightarrow x = y = 0 \Rightarrow y\ R_3\ x$, $\forall\ x, y \in \mathbb{Z}$. Transitive, since $x\ R_3\ y$ and $y\ R_3\ z \Rightarrow x = y = z = 0 \Rightarrow x\ R_3\ z$, $\forall\ x, y, z \in \mathbb{Z}$. Not reflexive, since e.g. 1 $\not R$ 1.
(c) Suppose there is $x_0 \in S$ such that $x_0\ \not R\ y$, for any $y \in S$. Then the argument given does not prove $x_0\ R\ x_0$ (see (b) above, with $x_0 = 1$ and $S = \mathbb{Z}$).

Problems 2.2a

1 (a) All the residents of a given street; ... street.
(b) All the residents born in a given calender year; ... year (in which at least one resident was born).
(c) All the residents with a birthday on a given day of the year; ... day of the year (on which at least one resident has a birthday).
(d) All the animals of a given species; ... species present in the zoo.
(e) All the paintings by a given artist; ... each artist represented in the National Gallery.
(f) All the students enrolled on a given degree course; ... each degree offered at the university.

2 (a) $E_1 = \overline{1} = \{2r + 1: r \in \mathbb{Z}\}$; $E_2 = \overline{2} = \{2r: r \in \mathbb{Z}\}$.
  (b) $\overline{0} = \{0\}$, $\overline{1} = \{1, -1\}$, $\overline{2} = \{2, -2\}$, ..., $\overline{r} = \{r, -r\}$, ...

3 (a) $\overline{(1, 2)} = \{..., (-1, 0), (0, 1), (1, 2), (2, 3), (3, 4), ...\}$
  $= \{(a, a + 1): a \in \mathbb{Z}\}$.
  (b) $\overline{(0, -3)} = \{..., (-2, -5), (-1, -4), (0, -3), (1, -2), ...,\} =$
  $\{a, a-3): a \in \mathbb{Z}\}$.
  (c) $\overline{(a, b)} = \{ ..., (a-1, b-1), (a, b), (a+1, b+1), (a+2, b+2), ...\}$
  $= \{(a+m, b+m): m \in \mathbb{Z}\}$.

The set of points with coordinates (x, y), where $(x, y) \in \overline{(a, b)}$, are all the points of the line $y - x = b - a$ in $\mathbb{Z} \times \mathbb{Z}$ (i.e. the line through the point (a, b) with gradient 1).

4 (a) $\overline{1} = \{1, 4, 9, 25, ...\} = \{r^2: r \in \mathbb{Z}\}$.
  (b) $\overline{2} = \{2, 8, 18, 50, ...\} = \{2r^2: r \in \mathbb{Z}\}$.
  (c) $\overline{p} = \{p, 4p, 9p, 25p, ...\} = \{pr^2: r \in \mathbb{Z}\}$.

5 (a) $R(2) = \emptyset$; $R(3) = \{2\}$; $R(4) = \{2, 3\}$; $R(6) = \{2, 3, 4\}$.
  (b) $a = 2$. (c) e.g. $a = 3$, $b = 4$. (d) Show that R is not reflexive or that R is not symmetric. No; this is proved in result 2.2.3 (page 43).

Problems 2.3a

1 1:

X_1	1	3	5	7	9	11	13	...
X_2	2	4	6	8	10	12	14	...

2:

X_1	X_2	X_3	X_4	X_5	X_6	X_7	X_8	X_9	X_{10}
1	2	3	4	5	6	7	8	9	10
11	12	13	14	15	16	17	18	19	20
21	22	23	24	25	26	27	28	29	30
⋮	⋮	⋮	⋮	⋮	⋮	⋮	⋮	⋮	⋮

2 1: $x \sim y \Leftrightarrow x$ and y have the same parity (or $x - y$ is even).
  2: $x \sim y \Leftrightarrow x$ and y terminate in the same digit (or $x - y$ is divisible by 10).
  3: $x \sim y \Leftrightarrow x$ and y have the same number of digits.

3 (i) By the definition of E_k, $k \in E_k$ and hence the cells of Σ are non-empty.

   (ii) Given $x \in \mathbb{R}$, let $k = [x]$ (i.e. the greatest integer $\leq x$, see problem 2b, 3). Then $k \leq x < k + 1$ and hence $x \in E_k$. Thus every element of $\mathbb{R}$ belongs to a cell in Σ.

   (iii) Suppose E_k, E_l are distinct members of Σ. Then $l \neq k$. Assume $l < k$. Then $l + 1 \leq k$, since l, k are integers. Now $(x \in E_l, l \leq x < l + 1) \Rightarrow x < k \Rightarrow x \notin E_k$. Similarly, $y \in E_k \Rightarrow y \geq k \Rightarrow y \geq l + 1 \Rightarrow y \notin E_\ell$. Then $E_l \cap E_k = \varnothing$.

Together, (i), (ii) and (iii) show that Σ partitions $\mathbb{R}$. Define an equivalence relation $\sim$ by $x \sim y \Leftrightarrow [x] = [y]$.

4 Define $\sim$ by: $\left. \begin{array}{l} a \sim x \\ b \sim x \\ c \sim x \end{array} \right\}$, $\forall\, x \in \{a, b, c\}$, and $d \sim d$.

Problems 2.4a

1 (a) $\bar{0} = \{2k: k \in \mathbb{Z}\}$, $\bar{1} = \{1 + 2k: k \in \mathbb{Z}\}$
   (b) $\bar{0} = \{5k: k \in \mathbb{Z}\}$, $\bar{1} = \{1 + 5k: k \in \mathbb{Z}\}$, $\bar{2} = \{2 + 5k: k \in \mathbb{Z}\}$,
        $\bar{3} = \{3 + 5k; k \in \mathbb{Z}\}$, $\bar{4} = \{4 + 5k: k \in \mathbb{Z}\}$.

2 (a) $\{a + 3k: k \in \mathbb{Z}\}$; $\{a + mk: k \in \mathbb{Z}\}$.

3 (a) (i) 2; (ii) 1; (iii) 0; (iv) 1; (v) 8.
   (b) (i) $\{2 + 3k: k \in \mathbb{Z}\}$; (ii) $\{1 + 5k: k \in \mathbb{Z}\}$; (iii) $\{3k: k \in \mathbb{Z}\}$;
        (iv) $\{1 + 2k: k \in \mathbb{Z}\}$; (v) $\{8 + 10k: k \in \mathbb{Z}\}$.

4 $x_0 = 7$; $y \equiv 7 \pmod{9} \Rightarrow y = 7 + 9k$, for some $k \in \mathbb{Z}$.
   Hence $2y = 14 + 9.2k \Rightarrow 2y \equiv 14 \pmod 9 \Rightarrow 2y \equiv 5 \pmod 9$.

5 No integer of the form $1 + 10k$ is divisible by 5.

6 $a \equiv b \pmod m \Rightarrow a = b + mk$, for some $k \in \mathbb{Z}$.

   (a) $a = b + mk \Rightarrow a + c = b + c + mk \Rightarrow a + c \equiv b + c \pmod m$;
   (b) $a = b + mk \Rightarrow ac = bc + m(kc) \Rightarrow ac \equiv bc \pmod m$.

Problems 2.5a

1 See Fig. 2.7.

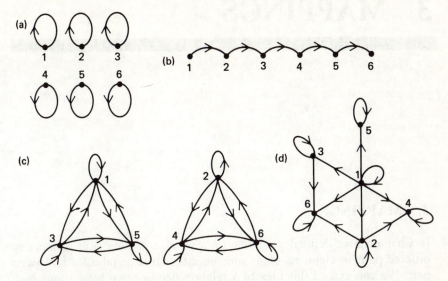

Fig. 2.7

2 (a) R_1, R_3, R_4; (b) R_1, R_3; (c) R_1, R_3, R_4.

R reflexive $\Leftrightarrow$ there is a loop at each point.

R symmetric $\Leftrightarrow$ no pair is connected by just one directed arc; i.e. whenever there is an arc from a to b, there is also an arc from b to a

R transitive $\Leftrightarrow$ whenever there is an arc from a to b and an arc from b to c, then there is an arc from a to c.

3 (a) R_1 and R_3; (b) R_2 and R_3; (c) R_1 and R_3.

3 MAPPINGS

3.1 MAPPINGS

In Chapter 2 we defined a relation on a set A as a statement connecting ordered pairs of elements of A, which is either true or false for any given pair. We can extend this idea to a relation *between* two sets A and B as a statement connecting ordered pairs of elements $a \in A$ and $b \in B$ which is either true or false for any given pair (a, b). As in section 2.5, we can illustrate such a relation by a *directed graph* with arcs (a, b), for all $a \in A$, $b \in B$ for which the statement is true. The first set A is called the *domain* of the relation, and the second set B is called the *codomain*.

Mappings

Let X, Y be non-empty sets and α be a relation between X and Y. Then α is said to be a *mapping* if *every* element in the domain X is related to *one* and *only one* element in the codomain Y.

Examples 3.1.1

1. Let X be the set of surnames in the London telephone directory and Y be the set of letters of the alphabet. Let α be the relation between X and Y which connects each surname in X with its initial letter.

   Thus Arbuthnot α a and Williams α w, for example. Given a surname, it has a *unique* initial letter and hence each element of X is related to *exactly one* element of Y; therefore α is a *mapping*.

   Note: The definition of mapping does not say that different members of X have to be related to *different* members of Y. Clearly, in this example many different members of X are related to the *same* member of Y.

2. Suppose a relation β is defined on $\mathbb{Z}^+$ by $a \beta b \Leftrightarrow b = 2a$. Then the arc set A of the directed graph of β is given by $A = \{(1, 2), (2, 4), (3, 6), (4, 8), \ldots\}$. Every $a \in \mathbb{Z}^+$ occurs as the first element of one of the ordered pairs *exactly once*. Hence every $a \in \mathbb{Z}^+$ is related to exactly one $b \in \mathbb{Z}^+$, therefore α is a *mapping*.

Note: The definition of a *mapping* does not require that every element of the *codomain* should occur as the second member of an ordered pair in the graph.

3. *Counter-example*: Taking the universal set as the set of all living people, let M be the subset of all males and F the subset of all females. Define the relation **R** between M and F by m **R** $f \Leftrightarrow m$ is the brother of f. Then **R** is not a mapping because:

   (a) not every man has a sister and hence there are some elements $m \in M$ which are not related to any element of F;
   (b) some men have more than one sister and hence some elements of M are related to more than one element of F.

Note: Either of the reasons (a) and (b) is sufficient to prove that **R** is *not* a mapping.

A relation α which is a mapping between a set X and a set Y is called simply a *mapping* of the set X *into* the set Y. We write

$$\alpha: X \rightarrow Y \qquad \text{or } X \xrightarrow{\alpha} Y.$$

The 'departure' set, X, is called the *domain* of the mapping and the 'arrival' set, Y, is called the *codomain*.

Images

Let $\alpha: X \rightarrow Y$ be a mapping. If $x \in X$ is related to $y \in Y$, y is called the *image of x under* α and we write $y = \alpha(x)$. (This relation will be familiar from the function notation in calculus. However, not all the relations referred to as 'functions' in calculus are *mappings* of $\mathbb{R}$ into $\mathbb{R}$ by our definition; see problems 3.1a below.) If $y = \alpha(x)$, then x is called the *pre-image* or *ancestor* of y.

Image set of a mapping

Let $\alpha: X \rightarrow Y$ be a mapping. The subset of Y containing all the elements which are *images* under α is called the *image set* of α, denoted by Im α.

This set is also sometimes called the *range* of α, a term which comes from calculus, where the image set of a function is often called its *range*.

Examples 3.1.2

1. Under the mapping $\beta: \mathbb{Z}^+ \to \mathbb{Z}^+$ defined in example 3.1.1, 2, above, Im β is the set of even positive integers. In this case Im β is a *proper* subset of the codomain.

2. Let $\alpha: \mathbb{Z} \to \mathbb{Z}$ be the mapping defined by $\alpha(r) = r+3$. To find Im α, we must ask: which members of the *codomain* $\mathbb{Z}$ can be written in the form $r+3$ for some r belonging to the *domain* $\mathbb{Z}$? Clearly every integer m can be written in the form $r+3$ for $r = m-3 \in \mathbb{Z}$, and hence Im α = $\mathbb{Z}$.

3. Let $\alpha: \mathbb{R} \to \mathbb{R}$ be defined by $\alpha(x) = x^2 - 2x$. To find Im α, let y be any element of the *codomain* $\mathbb{R}$. We ask: for what values of y can we find $x \in \mathbb{R}$ (the *domain* of α) such that $\alpha(x) = x^2 - 2x = y$? Now $x^2 - 2x = y \Rightarrow (x - 1)^2 = y + 1$. Hence a solution for x exists if and only if $y + 1 \geq 0$. Thus Im α = $\{y \in \mathbb{R}: y \geq -1\}$.

   In the sketch graph of $\alpha(x) = x^2 - 2x$ (Fig. 3.1), the x-axis is representing the *domain* of α, the complete y-axis represents the *codomain* and Im α is the set of y-coordinates of points of the graph.

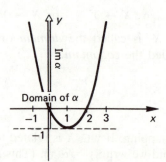

Fig. 3.1

Equality of mappings

Let $\alpha: A \to B$ and $\beta: X \to Y$ be mappings. Then we say that α and β are equal (α = β), if and only if they satisfy the following three conditions:

   (i) $A = X$;   (ii) $B = Y$;   (iii) $\alpha(x) = \beta(x)$, $\forall x \in A$.

Thus we are saying that $\alpha = \beta$ if and only if α and β have the *same domain* and the *same codomain* and if they map each element of their common domain into the *same image*.

PROBLEMS 3.1a

1 Each part of Fig. 3.2 below illustrates a relation between the set $S = \{1, 2, 3\}$ and the set $T = \{a, b\}$. State which of the relations illustrated are *not* mappings, giving a reason in each case.

(a)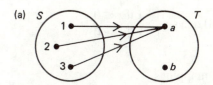

(b)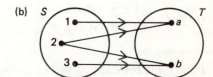

(c)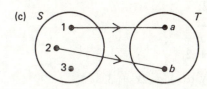

(d)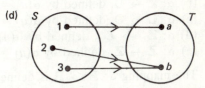

Fig. 3.2

2 (a) The following are relations on the set of real numbers $\mathbb{R}$. Which are *mappings* of $\mathbb{R}$ into $\mathbb{R}$? Explain briefly why each of the others are not.

   (i) $\alpha: x \mapsto y \Leftrightarrow x + y \le 6$;   (ii) $\alpha: x \mapsto y \Leftrightarrow y = x^2$;
   (iii) $\alpha: x \mapsto y \Leftrightarrow y^2 = x$;   (iv) $\alpha: x \mapsto y \Leftrightarrow xy = 4$;
   (v) $\alpha: x \mapsto y \Leftrightarrow x^2 + y^2 = 1$;
   (vi) $\alpha: x \mapsto y \Leftrightarrow x^2 + y^2 = 1, y \ge 0$;
   (vii) $\alpha: x \mapsto y \Leftrightarrow y = \sin x$;   (viii) $\alpha: x \mapsto y \Leftrightarrow x = \sin y$;
   (ix) $\alpha: x \mapsto y \Leftrightarrow y = (x-1)(x-3)$;
   (x) $\alpha: x \mapsto y \Leftrightarrow x = (y-1)(y-3)$.

  (b) Give the image set, Im α, of each of the *mappings* in (a).

3 The following relations are not mappings of $\mathbb{R}$ into $\mathbb{R}$. In each case give the minimal subset of real numbers which should be excluded from the domain to make the relation a mapping.

(a) $\beta: x \mapsto \dfrac{2}{x+3}$;   (b) $\beta: x \mapsto \dfrac{1}{x^2-1}$;

(c) $\beta: x \mapsto \ln x$;   (d) $\beta: x \mapsto \tan x$;

(e) $\beta: x \mapsto \sqrt{x}$ (where $\sqrt{x}$ denotes the *positive* square root of x).

4 Decide whether any of the following relations on $\mathbb{R}$ are mappings of $\mathbb{R} \to \mathbb{R}$; if you think a relation is not a mapping explain briefly why not.

(a) $\alpha: x \mapsto \begin{cases} \dfrac{1}{1-x} & \text{if } x \le 1; \\ \dfrac{1}{x-1} & \text{if } x > 1; \end{cases}$

(b) $\alpha: x \mapsto \begin{cases} \dfrac{1}{1-x} & \text{if } x < 1; \\ 0 & \text{if } x = 1; \\ \dfrac{1}{x-1} & \text{if } x > 1; \end{cases}$

(c) $\alpha: x \mapsto \begin{cases} 1-x & \text{if } x \le 1; \\ 1+x & \text{if } x \ge 1; \end{cases}$

(d) $\alpha: x \mapsto \begin{cases} -x & \text{if } x \le 0; \\ x & \text{if } x \ge 0. \end{cases}$

5 Find Im α for each of the following mappings:

(a) $\alpha: \mathbb{Z}^+ \to \mathbb{Z}^+$, defined by $\alpha(n) = 2n + 1$;

(b) $\alpha: \mathbb{Z} \to \mathbb{Q}$, defined by $\alpha(a) = a/2$;

(c) $\alpha: \mathbb{Q} \to \mathbb{Q}$, defined by $\alpha(q) = q/2$;

(d) $\alpha: \mathbb{Z}^+ \to \mathbb{Z}^+$, defined by $\alpha(n) = n + 2$;

(e) $\alpha: \mathbb{Z} \to \mathbb{Z}$, defined by $\alpha(a) = a + 2$.

6 The mapping $\alpha: \mathbb{R} \to \mathbb{R}$ is defined by $\alpha(x) = 2x + 3$.

Find   (a) $\alpha(2x)$;   (b) $\alpha(x - 1)$;   (c) $\alpha\left(\dfrac{1}{x}\right)$;   (d) $\alpha(x^2)$.

PROBLEMS 3.1b

1 Let $S = \{1, 2, 3, 4\}$. Decide which of the following sets of ordered pairs are *not* the arc sets of a directed graph of a mapping of $S \to S$, giving a reason in each case.

(a) $\{(1, 2), (3, 4), (2, 1), (4, 3)\}$;   (b) $\{(4, 1), (2, 2), (3, 4)\}$;

(c) $\{(2, 2), (4, 4), (3, 1), (2, 3), (1, 4)\}$;

(d) $\{(1, 2), (2, 2), (3, 2), (4, 2)\}$.

2 (a) Let $S = \{1, 2, 3, 4\}$ and A be the arc set of a directed graph of a mapping $\alpha: S \to \mathbb{Z}^+$. How many distinct ordered pairs does A contain?

(b) Let $S = \{1, 2, 3, 4\}$, $X = \{x\}$, $Y = \{x, y\}$ and $Z = \{x, y, z\}$.

How many distinct mappings can be defined from S into (i) X; (ii) Y; (iii) Z?

3 (a) The mapping $\alpha: \mathbb{R} \to \mathbb{R}$ is defined by $\alpha(x) = x^2 - 2x$. Find

(i) $\alpha\left(\dfrac{x}{2}\right)$; (ii) $\alpha\left(\dfrac{1}{x}\right)$; (iii) $\alpha(x^2)$; (iv) $\alpha(1 + x)$.

(b) Give an example of a mapping $\beta; \mathbb{R} \to \mathbb{R}$ such that

(i) $\beta(x) = \beta(-x)$, $\forall\, x \in \mathbb{R}$; (ii) $\beta(x) = \beta\left(\dfrac{1}{x}\right)$, $\forall\, x \in \mathbb{R}, x \neq 0$.

4 Find the image set, Im α, of each of the following mappings, α:

(a) $\alpha: \mathbb{Z}^+ \to \mathbb{Z}^+$ defined by $\alpha(n) = 2n + 5$.
(b) $\alpha: \mathbb{Z} \to \mathbb{Z}$ defined by $\alpha(r) = 2r + 5$.
(c) $\alpha: \mathbb{Q} \to \mathbb{Q}$ defined by $\alpha(q) = \frac{1}{4}(q - 1)$.
(d) $\alpha: \mathbb{R} \to \mathbb{R}$ defined by $\alpha(x) = x^2 - 5$.
(e) $\alpha: \mathbb{R} \to \mathbb{R}$ defined by $\alpha(x) = (4 - x)(x - 2)$.
(f) $\alpha: \mathbb{R} \to \mathbb{R}$ defined by $\alpha(x) = x^3$.
(g) $\alpha: \mathbb{R} \to \mathbb{R}$ defined by $\alpha(x) = e^x$.
(h) $\alpha: \mathbb{R} \to \mathbb{R}$ defined by $\alpha(x) = 1/(x^2 + 1)$.

5 Let $X = \mathbb{Q} - \{1\}$, and $\alpha: X \to \mathbb{Q}$ be the mapping defined by $\alpha(a) = \dfrac{2a - 3}{a - 1}$. Show that there is just one rational number not in Im α and find it.

3.2 SPECIAL TYPES OF MAPPINGS

Surjective mappings

A mapping α is called *surjective* if every element in the codomain is in the image set of α. Thus

$$\alpha \text{ is surjective} \Leftrightarrow \text{Im } \alpha = \text{codomain of } \alpha.$$

If $\alpha: A \to B$ is a surjective mapping we say that α maps the set A *onto* the set B. For this reason a surjective mapping is sometimes called an 'onto' mapping.

The term *surjective mapping* is sometimes abbreviated to *surjection*.

Examples 3.2.1

1. Let $\alpha: \mathbb{Z}^+ \to \mathbb{Z}^+$ be defined by $\alpha(n) = n + 1$. Then 1 has no ancestor in $\mathbb{Z}^+$ and hence Im $\alpha = \{n \in \mathbb{Z}^+: n \geq 2\}$ and so α is not surjective.

2. Let $\beta: \mathbb{Z} \to \mathbb{Z}$ be defined by $\beta(n) = n + 1$. In this case Im $\beta = \mathbb{Z} = $ codomain of β and hence β is surjective.

Injective mappings

A mapping $\alpha\colon A \to B$ is called *injective* if *distinct* elements of A have *distinct* images in B. That is, $a_1 \neq a_2 \Rightarrow \alpha(a_1) \neq \alpha(a_2)$. It can be seen from that an *injective* mapping $\alpha\colon A \to B$ is a 'one-to-one matching' or 'one-to-one correspondence' between the elements of A and the elements of *Im* α. For this reason, an injective mapping is sometimes called a 'one-to-one' mapping.

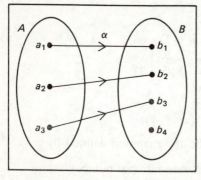 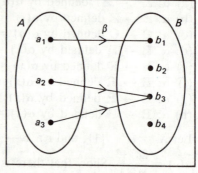

(a) $\alpha\colon A \to B$ is injective;   (b) $\beta\colon A \to B$ is *not* injective.

Fig. 3.3

To prove that a mapping $\alpha\colon A \to B$ is *injective*, we must show that for any pair of elements $a_1, a_2 \in A$,

$$\alpha(a_1) = \alpha(a_2) \Rightarrow a_1 = a_2.$$

That is, if two *images* coincide, then their ancestors must be the same element of A. An *injective* mapping is also called an *injection*.

Examples 3.2.2

1. Let $\beta\colon \mathbb{Z} \to \mathbb{Z}$ be defined by $\beta(a) = 2a + 3$. Then $\beta(a_1) = \beta(a_2) \Rightarrow 2a_1 + 3 = 2a_2 + 3 \Rightarrow a_1 = a_2$. Hence β is *injective*.

2. Let $\gamma\colon \mathbb{R} \to \mathbb{R}$ be defined by $\gamma(x) = x^2 - 4x$. Then

$$\gamma(x_1) = \gamma(x_2) \Rightarrow x_1^2 - 4x_1 = x_2^2 - 4x_2;$$
$$\Rightarrow x_1^2 - x_2^2 - 4(x_1 - x_2) = 0;$$
$$\Rightarrow (x_1 - x_2)(x_1 + x_2 - 4) = 0;$$
$$\Rightarrow x_1 = x_2 \quad or \quad x_1 = 4 - x_2.$$

Thus γ is *not* injective.
Figure 3.4 illustrates $\gamma(x_1) = \gamma(x_2)$ but $x_1 \neq x_2$.

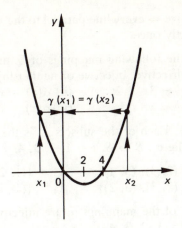

Fig. 3.4

Bijective mappings

A mapping $\alpha: A \to B$ which is both *surjective* and *injective* is called a *bijective mapping* (or a bijection) of A onto B. Combining the conditions for α to be both surjective and injective, we see that a *bijection* is a one-to-one correspondence between (all) the elements of A and (all) the elements of B.

To prove that a mapping $\alpha: A \to B$ is *bijective* we must show that:

(i) $\text{Im } \alpha = B$; (ii) $\alpha(a_1) = \alpha(a_2) \Rightarrow a_1 = a_2$.

PROBLEMS 3.2a

1 Decide whether each of the following mappings is injective, surjective or bijective.

(a) $\alpha: \mathbb{Z}^+ \to \mathbb{Z}^+$ defined by $\alpha(n) = n + 3$;
(b) $\alpha; \mathbb{Z} \to \mathbb{Z}$ defined by $\alpha(r) = -r$;
(c) $\alpha: \mathbb{Z} \to \mathbb{Z}$ defined by $\alpha(r) = 3r - 1$;
(d) $\alpha: \mathbb{Q} \to \mathbb{Q}$ defined by $\alpha(q) = 3q - 1$;
(e) $\alpha: \mathbb{Z}^+ \to \mathbb{Z}^+$ defined by $\alpha(n) = n^2$;
(f) $\alpha: \mathbb{Z} \to \mathbb{Z}$ defined by $\alpha(x) = x^2$.

2 (a) Let α be a mapping of $\mathbb{R}$ into $\mathbb{R}$. Show that
  (i) α is surjective $\Leftrightarrow$ every line parallel to the x-axis meets the graph of α at least once.
  (ii) α is injective $\Leftrightarrow$ every line parallel to the x-axis meets the graph of α at most once.

(iii) α is bijective $\Leftrightarrow$ every line parallel to the x-axis meets the graph of α exactly once.

(b) For each of the following mappings of $\mathbb{R}$ into $\mathbb{R}$, say whether it is injective, surjective, bijective or neither injective nor surjective:
(i) $\alpha(x) = x^2 - 3x + 2$;   (ii) $\alpha(x) = x^3$;
(iii) $\alpha(x) = x^3 - x$;   (iv) $\alpha(x) = \exp(x)$.

3 Let $S = \{1, 2, 3\}$. Each of the subsets A_i is the arc set of a directed graph of a mapping $\alpha_i: S \to S$, $i = 1, 2, 3, 4$.

$A_1 = \{(1, 1), (2, 3), (3, 1)\}$; $A_2 = \{(1, 2), (2, 3), (3, 1)\}$;
$A_3 = \{(1, 1), (2, 3), (3, 2)\}$; $A_4 = \{(1, 3), (2, 3), (3, 3)\}$.

(a) Decide which of the mappings α_i are injective.
(b) Find Im α_i, $i = 1, 2, 3, 4$.
(c) Which mappings α_i are surjective?
(d) Is it possible to define a mapping $\alpha: S \to S$ which is injective but not surjective? Surjective but not injective?

4 Prove that the mapping $\alpha: \mathbb{R} - \{\frac{1}{2}\} \to \mathbb{R}$, defined by $\alpha(x) = \dfrac{x+1}{2x-1}$, is injective but not surjective.

5 Let S be a finite set of n elements and $\alpha: S \to S$ be a mapping. Show that α is injective if and only if the number of elements in Im α is n. Deduce that α is injective if and only if it is surjective. Give a counter-example to prove that this result does not hold for a mapping β of an *infinite* set into itself.

PROBLEMS 3.2b

1 The mappings α, $\beta: \mathbb{Z} \to \mathbb{Z}$ are defined by:

$$\alpha(a) = \begin{cases} a+1 \text{ if } a \text{ is even,} \\ 2a \text{ if } a \text{ is odd;} \end{cases} \qquad \beta(a) = \begin{cases} a+2 \text{ if } a \text{ is even,} \\ \frac{1}{2}(a+1) \text{ if } a \text{ is odd.} \end{cases}$$

Decide, giving reasons, whether either of the mappings α, β is (a) injective (b) surjective.

2 Mappings $\alpha_i: \mathbb{Z} \times \mathbb{Z} \to \mathbb{Z}$, $i = 1, 2, 3, 4$, are defined by:

$$\alpha_1: (a, b) \mapsto a; \qquad \alpha_2: (a, b) \mapsto a + b;$$
$$\alpha_3: (a, b) \mapsto 2b; \qquad \alpha_4: (a, b) \mapsto ab.$$

Decide, giving reasons, which of the mappings α_i are surjective. Prove, by giving suitable counter-examples, that none of the mappings α_i, $i = 1, 2, 3, 4$, is injective.

3 Let X, Y be finite sets with $|X| = m$ and $|Y| = n$.

   (a) Find a formula for the number of different mappings that it would be possible to define from X into Y.

   (b) If $m \leq n$, find a formula for the number of injections that could be defined from X into Y.

   (c) What can be said about m and n if it is possible to define (a) a surjective (b) a bijective mapping of X into Y?

3.3 COMPOSITION OF MAPPINGS

Let A, B, C be non-empty sets and α: $A \to B$, β: $B \to C$ be mappings. We define a mapping γ: $A \to C$ by the rule

$$\gamma(a) = \beta[\alpha(a)].$$

Figure 3.5 illustrates the 'path' of an element $a \in A$.
If $\alpha(a) = b \in B$ and $\beta(b) = c \in C$ then $\gamma(a) = \beta[\alpha(a)] = \beta[b] = c$.

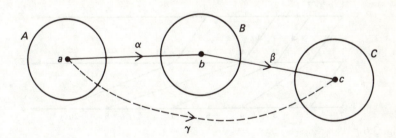

Fig. 3.5

Composite mapping

The mapping γ: $A \to C$ defined above is called the *composite mapping* (or *composition*) of α by β. This is written

$$\gamma = \beta \circ \alpha.$$

Note carefully the order in which the mappings are written; to find $(\beta \circ \alpha)(a)$, we first find $\alpha(a) = b$, say, and secondly find $\beta(b)$, so that the mapping written *last* is actually the *first* mapping to be used, that is, we work from *right* to *left*.

Examples 3.3.1

1. Let $A = \{1, 2, 3, 4\}$, $B = \{x, y, z, w\}$, $C = \{a, b, c\}$.
   Mappings $\alpha: A \rightarrow B$ and $\beta: B \rightarrow C$ are defined by:

$$\alpha(1) = x, \quad \alpha(2) = y, \quad \alpha(3) = x, \quad \alpha(4) = z;$$
$$\beta(x) = b, \quad \beta(y) = a, \quad \beta(z) = b, \quad \beta(w) = c.$$

We define the compositive mapping $\beta \circ \alpha: A \rightarrow C$.

Now, $\beta \circ \alpha: 1 \overset{\alpha}{\mapsto} x \overset{\beta}{\mapsto} b$ so that $(\beta \circ \alpha)(1) = b$;
$\qquad\qquad 2 \mapsto y \mapsto a \qquad\qquad\qquad (\beta \circ \alpha)(2) = a$;
$\qquad\qquad 3 \mapsto x \mapsto b \qquad\qquad\qquad (\beta \circ \alpha)(3) = b$;
$\qquad\qquad 4 \mapsto z \mapsto b \qquad\qquad\qquad (\beta \circ \alpha)(4) = b$.

2. Let $\alpha: \mathbb{Z}^+ \rightarrow \mathbb{Z}^+$ and $\beta: \mathbb{Z}^+ \rightarrow \mathbb{Z}^+$ be defined by $\alpha(n) = n + 1$ and $\beta(n) = 2n$. We define the composite mappings (a) $\beta \circ \alpha$ and (b) $\alpha \circ \beta$.

(a) Figure 3.6 illustrates the mapping $\beta \circ \alpha: \mathbb{Z}^+ \rightarrow \mathbb{Z}^+$, defined by:

$$(\beta \circ \alpha)(n) = \beta[\alpha(n)]$$
$$= \beta[n + 1]$$
$$= 2(n + 1).$$

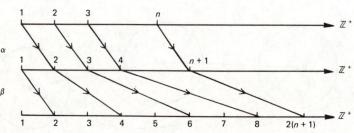

Fig. 3.6

(b) Figure 3.7 illustrates $\alpha \circ \beta: \mathbb{Z}^+ \rightarrow \mathbb{Z}^+$, defined by

$$(\alpha \circ \beta)(n) = \alpha(\beta(n))$$
$$= \alpha(2n)$$
$$= 2n + 1.$$

Notice that $(\alpha \circ \beta)(n) \neq (\beta \circ \alpha)(n)$ and hence $\alpha \circ \beta \neq \beta \circ \alpha$.

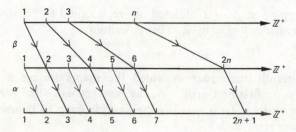

Fig. 3.7

Domains and image sets

Let A, B, C be non-empty sets and $\alpha: A \to B$, $\beta: B \to C$. Then A is the domain of α and B is the codomain of α. In future we will write $A = \text{Dom } \alpha$, $B = \text{Codom } \alpha$.

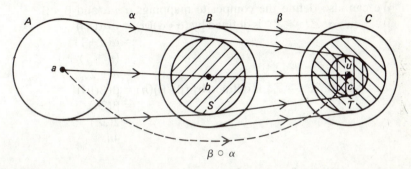

$\beta \circ \alpha$

Fig. 3.8

In Fig. 3.8:

$$\text{Dom } \alpha = A, \qquad \text{Codom } \alpha = B. \qquad \text{Im } \alpha = S.$$
$$\text{Dom } \beta = B, \qquad \text{Codom } \beta = C. \qquad \text{Im } \beta = T.$$
$$\text{Dom } \beta \circ \alpha = A, \quad \text{Codom } \beta \circ \alpha = C. \quad \text{Im } \beta \circ \alpha = U.$$

Existence of $\beta \circ \alpha$ and $\alpha \circ \beta$

Let α, β be mappings.
The mapping $\beta \circ \alpha$ is *only* defined when

$$\text{Im } \alpha \subseteq \text{Dom } \beta.$$

Similarly, the mapping $\alpha \circ \beta$ is *only* defined when

$$\text{Im } \beta \subseteq \text{Dom } \alpha.$$

Thus $\beta \circ \alpha$ and $\alpha \circ \beta$ can *both* be defined *only* when

$$\alpha: A \to B \text{ and } \beta: B \to A, \text{ for some sets } A \text{ and } B.$$

If $\alpha: A \to A$, then the composite mapping $\alpha \circ \alpha: A \to A$ can always be defined.

Examples 3.3.2

1. In example 3.3.1, 1 above, we defined the composite mapping $\beta \circ \alpha: A \to C$. Here, $\text{Im } \alpha = \{x, y, z\} \subseteq B$ and $\text{Dom } \beta = B$.

69

However, the composite mapping $\alpha \circ \beta$ cannot be defined because Im $\beta = C$ and α is not defined on the elements of C.
Similarly we cannot define $\alpha \circ \alpha$ or $\beta \circ \beta$.

2. In example 3.3.1, 2 above, $\alpha: \mathbb{Z}^+ \to \mathbb{Z}^+$ and $\beta: \mathbb{Z}^+ \to \mathbb{Z}^+$. Thus Im $\alpha \subseteq$ Dom β and Im $\beta \subseteq$ Dom α and so we were able to define both $\beta \circ \alpha$ and $\alpha \circ \beta$.

We can also define the composite mappings $\alpha \circ \alpha$ and $\beta \circ \beta$:

$$\alpha \circ \alpha: \mathbb{Z}^+ \to \mathbb{Z}^+ \text{ is defined by } (\alpha \circ \alpha)(n) = \alpha(\alpha(n))$$
$$= \alpha(n+1)$$
$$= (n+1)+1$$
$$= n+2;$$

$$\beta \circ \beta: \mathbb{Z}^+ \to \mathbb{Z}^+ \text{ is defined by } (\beta \circ \beta)(n) = \beta(\beta(n))$$
$$= \beta(2n)$$
$$= 2(2n)$$
$$= 4n.$$

Special properties of composite mappings

The following results show how the injective and surjective properties of mappings and of their composite mappings are related. Note the *strategy* employed in the proofs of the results; it is exactly the same as that employed in the numerical examples in the previous section.

THEOREM 3.3.1   Let $\alpha: A \to B$ and $\beta: B \to C$ be bijective. Then $\beta \circ \alpha: A \to C$ is also a bijection.

Proof

(i) To show that $\beta \circ \alpha$ is *injective*.
Suppose $(\beta \circ \alpha)(a_1) = (\beta \circ \alpha)(a_2)$, for some $a_1, a_2 \in A$.
Then $\beta(\alpha(a_1)) = \beta(\alpha(a_2))$
$\Rightarrow \alpha(a_1) = \alpha(a_2)$ since β is injective
$\Rightarrow \quad a_1 = \quad a_2$ since α is injective.

(ii) To show that $\beta \circ \alpha$ is *surjective*.
Let $c \in C$ be any element of Codom $\beta \circ \alpha$.
Then since β is surjective, we can find $b \in B$ such that $\beta(b) = c$.
But since α is surjective, we can find $a \in A$ such that $\alpha(a) = b$.
Then $(\beta \circ \alpha)(a) = \beta(\alpha(a)) = \beta(b) = c$.
Hence given any element $c \in C$, there exists an element $a \in A$ such that $(\beta \circ \alpha)(a) = c$. Thus $\beta \circ \alpha$ is surjective. ■

RESULT 3.3.2   Let $\alpha: A \to B$ and $\beta: B \to C$ be mappings such that $\beta \circ \alpha: A \to C$ is injective. Then α is injective.

Proof Suppose $\alpha(a_1) = \alpha(a_2)$, for some $a_1, a_2 \in A$.
Then $\beta(\alpha(a_1)) = \beta(\alpha(a_2))$;
$\Rightarrow (\beta \circ \alpha)(a_1) = (\beta \circ \alpha)(a_2)$
$\Rightarrow a_1 = a_2$, since $\beta \circ \alpha$ is injective.
Hence α is injective. ■

RESULT 3.3.3 Let $\alpha: A \to B$ and $\beta: B \to C$ be mappings such that $\beta \circ \alpha: A \to C$ is surjective. Then β is surjective.

Proof Let $c \in C$ be any element in Codom β.
Then since $\beta \circ \alpha$ is surjective, we can find $a \in A$ such that $(\beta \circ \alpha)(a) = c$.
But $(\beta \circ \alpha)(a) = c \Rightarrow \beta(\alpha(a)) = c$.
Let $\alpha(a) = b$. Then $\beta(b) = c$.
Hence given any element $c \in C$, we can find an element $b \in B$ such that $\beta(b) = c$. Hence β is surjective. ■

PROBLEMS 3.3a

1 For each of the following pairs of mappings $\alpha, \beta: \mathbb{R} \to \mathbb{R}$ find $\beta \circ \alpha$, $\alpha \circ \beta$, $\alpha \circ \alpha$ and $\beta \circ \beta$.

(a) $\alpha(x) = \sin x$; $\beta(x) = 2x + 1$; (b) $\alpha(x) = (x+1)/3$; $\beta(x) = 3x - 1$;
(c) $\alpha(x) = x^2$; $\beta(x) = x - 2$.

2 Let $\alpha, \beta: \mathbb{R} \to \mathbb{R}$ be defined by $\alpha(x) = |x|$, $\beta(x) = \sin x$, and $\gamma: \mathbb{R}^+ \to \mathbb{R}$ be defined by $\gamma(x) = \ln x$, where $\mathbb{R}^+ = \{x \in \mathbb{R}: x > 0\}$.
Consider each of the six composite mappings $\alpha \circ \beta$, $\beta \circ \alpha$, $\alpha \circ \gamma$, $\gamma \circ \alpha$, $\beta \circ \gamma$, $\gamma \circ \beta$; define it if possible, or state why it cannot be defined.

3 Let $\alpha: A \to A$ be any mapping and let $\iota: A \to A$ be defined by $\iota(a) = a$, $\forall\, a \in A$. Show that (a) $\iota \circ \alpha = \alpha$; and (b) $\alpha \circ \iota = \alpha$.

4 State the *converse* of theorem 3.3.1.
Let $\alpha: \mathbb{Z} \to \mathbb{Z} \times \mathbb{Z}$ be defined by $\alpha(a) = (a + 3, 1)$ and $\beta: \mathbb{Z} \times \mathbb{Z} \to \mathbb{Z}$ be defined by $\beta(a, b) = a + b$.
(a) Show that α is injective but not surjective.
(b) Show that β is surjective but not injective.
(c) Define $\beta \circ \alpha$ and show that it is bijective.
Deduce that the converse of theorem 3.3.1 is *false*. Define also $\alpha \circ \beta$. Is this mapping bijective?

PROBLEMS 3.3b

1 For each of the following pairs of mappings α, β: $\mathbb{R} \to \mathbb{R}$, define $\beta \circ \alpha$, $\alpha \circ \beta$, $\alpha \circ \alpha$ and $\beta \circ \beta$.

(a) $\alpha(x) = 1 + \cos x$; $\beta(x) = 3x - 1$;
(b) $\alpha(x) = x^2 + 3$; $\beta(x) = 2x + 1$;
(c) $\alpha(x) = \exp(x)$; $\beta(x) = 2x$.

2 Let α: $\mathbb{Q} \to \mathbb{Q}$ be defined by $\alpha(q) = 3q - 2$. Give an example of a mapping β such that $\beta \circ \alpha = \alpha \circ \beta$ and $\beta(q) \neq q$, $\forall\ q \in \mathbb{Q}$.

3 Let α, β: $\mathbb{Z} \to \mathbb{Z}$ be defined by $\alpha(a) = 2a$ and

$$\beta(a) = \begin{cases} a/2 + 1, & \text{when } a \text{ is even,} \\ (a + 1)/2, & \text{when } a \text{ is odd.} \end{cases}$$

Show that neither α nor β are bijective, but that $\beta \circ \alpha$ is bijective.

4 Let α, β: $A \to B$ and γ: $B \to C$ be mappings such that $\gamma \circ \alpha = \gamma \circ \beta$ and γ is injective. Prove that $\alpha = \beta$. Let $A = B = C = \{1, 2, 3\}$. Define mappings α, β, γ: $A \to A$ such that $\alpha \neq \beta$, but $\gamma \circ \alpha = \gamma \circ \beta$. Deduce that $\gamma \circ \alpha = \gamma \circ \beta \not\Rightarrow \alpha = \beta$ if γ is *not* injective.

3.4 THE INVERSE OF A MAPPING

The identity map

Given a non-empty set A, we can define a mapping ι_A: $A \to A$ by the rule

$$\iota_A\ (a), \qquad a \in A.$$

Then ι_A is called the *identity mapping on A*. It maps every element of A onto itself, and hence it is easy to show that:

RESULT 3.4.1 For any set A, the identity map ι_A: $A \to A$ is bijective.

In some respects, the role of the identity map in composition of mappings is similar to the role of the number 1 in multiplication of numbers.

> RESULT 3.4.2   Let $\alpha: A \rightarrow B$. Then
>
>   (a) $\iota_B \circ \alpha = \alpha$;   (b) $\alpha \circ \iota_A = \alpha$.

Proof

(a) We first note that both $\iota_B \circ \alpha$ and α have domain A and codomain B.

Also, for every $a \in A$,

$$(\iota_B \circ \alpha)(a) = \iota_B(\alpha(a))$$
$$= \alpha(a), \qquad \text{by definition of } \iota_B.$$

Hence $\iota_B \circ \alpha = \alpha$.

(b) Again, $\alpha \circ \iota_A$ has domain A and codomain B.

Also, for every $a \in A$,

$$(\alpha \circ \iota_A(a) = \alpha(\iota_A(a))$$
$$= \alpha(a), \qquad \text{by definition of } \iota_A.$$

Hence $\alpha \circ \iota_A = \alpha$. ∎

The existence of the inverse of a mapping

Let $\alpha: A \rightarrow B$ be a mapping. Then we can define a relation β between the sets B and A by

$$\beta(b) = a \Leftrightarrow \alpha(a) = b.$$

Examples 3.4.1

1. Let $A = \{a_1, a_2, a_3\}$, $B = \{b_1, b_2, b_3\}$ and $\alpha: A \rightarrow B$ be defined by $\alpha(a_i) = b_i$, $i = 1, 2, 3$; see Fig. 3.9.

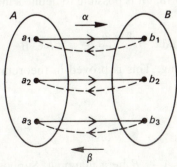

Here α is a *bijective* mapping (or one-to-one correspondence between the elements of A and B). Every element $b \in B$ is related to exactly one element of A, and hence β is a *mapping*.

Fig. 3.9

2. Let $A = \{a_1, a_2, a_3\}$, $B = \{b_1, b_2\}$ and $\alpha: A \to B$ be defined by
$\alpha(a_1) = b_1$, $\alpha(a_2) = \alpha(a_3) = b_2$; see Fig. 3.10.

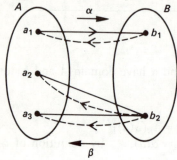

Here α is surjective but not injective. Since $b_2 \in B$ is related to *more* than one element of A, β is *not* a mapping.

Fig. 3.10

3. Let $A = \{a_1, a_2\}$, $B = \{b_1, b_2, b_3\}$ and $\alpha: A \to B$ be defined by
$\alpha(a_1) = b$, $\alpha(a_2) = b_2$; see Fig. 3.11.

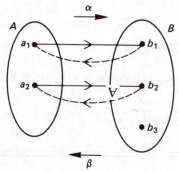

Here α is injective but not surjective. Since $b_3 \in B$ is not related to any element of A, β is *not* a mapping.

Fig. 3.11

From examples 3.4.1, we can suspect that it is possible to define a mapping
$\beta: B \to A$ such that

$$a \overset{\alpha}{\mapsto} b \overset{\beta}{\mapsto} a, \quad \forall\, a \in A, \quad \text{and} \quad b \overset{\beta}{\mapsto} a \overset{\alpha}{\mapsto} b, \quad \forall\, b \in B$$

if and only if α is a bijective mapping. This is proved in theorem 3.4.3, below.

The inverse of a mapping

Let A, B be non-empty sets and $\alpha: A \to B$ be a mapping. Suppose that
the relation β between the sets B and A defined by

$$\beta\ (b) = a \Leftrightarrow \alpha(a) = b$$

is a *mapping* of B into A. Then β is called the *inverse* mapping of α and is denoted by α^{-1}. The mapping α is said to be *invertible*.

THEOREM 3.4.3 Let $\alpha: A \to B$ be a mapping. Then the inverse mapping $\alpha^{-1}: B \to A$ exists if and only if α is bijective.

Note: The reader is advised to look at the diagrams in examples 3.4.1 above, while following the argument of this proof.

Proof Let $\beta: B \to A$ be the relation between B and A defined by $\beta(b) = a \Leftrightarrow \alpha(a) = b$. Then we need to show that β is a mapping $\Leftrightarrow \alpha$ is bijective.

   (i) Suppose α is bijective.

       Then since α is *surjective*, given any $b \in B$ we can find $a \in A$ such that $\alpha(a) = b$. Hence every $b \in B$ is related to at least one $a \in A$. But since α is also *injective*, given $b \in B$, there is *at most* one $a \in A$ such that $\alpha(a) = b$.

       Hence every $b \in B$ is related to *exactly one* $a \in A$ and thus β is a mapping.

   (ii) Suppose next that α is invertible so that $\beta = \alpha^{-1}$ is a mapping. Then given any $b \in B$, there exists *exactly one* element $a \in A$ such that $\alpha^{-1}(b) = a$. That is, given $b \in B$, there is *exactly one* element $a \in A$ such that $\alpha(a) = b$. Hence α is both injective and surjective and thus bijective. ∎

RESULT 3.4.4 Let $\alpha: A \to B$ be an invertible mapping. Then

      (a) $\alpha^{-1} \circ \alpha = \iota_A$;   (b) $\alpha \circ \alpha^{-1} = \iota_B$.

Proof

   (a) Let $a \in A$ and suppose $\alpha(a) = b$. Then $\alpha^{-1}(b) = a$, by definition of α^{-1}. Hence

$$(\alpha^{-1} \circ \alpha)(a) = \alpha^{-1}(\alpha(a)) = \alpha^{-1}(b) = a.$$

Clearly $\alpha^{-1} \circ \alpha: A \to A$ and hence $\alpha^{-1} \circ \alpha = \iota_A$.

(b) Let $b \in B$ and suppose $\alpha^{-1}(b) = a$. Then $\alpha(a) = b$, by definition of α^{-1}. Hence

$$(\alpha \circ \alpha^{-1})(b) = \alpha(\alpha^{-1}(b)) = \alpha(a) = b.$$

Clearly $\alpha \circ \alpha^{-1}: B \to B$ and hence $\alpha \circ \alpha^{-1} = \iota_B.$ ■

Suppose that $\alpha: A \to B$ is an invertible mapping. It is clear from the symmetry of the definition of α^{-1} (as well as from the symmetry of result 3.4.4) that α itself is the *inverse* mapping of α^{-1}. Figure 3.9 illustrates this. Thus we have:

RESULT 3.4.5 Let $\alpha: A \to B$ be an invertible mapping. Then $\alpha^{-1}: B \to A$ is also invertible and

$$(\alpha^{-1})^{-1} = \alpha.$$

PROBLEMS 3.4a

1 Decide which of the following mappings are invertible and either give the inverse mapping or explain why it does not exist.

(a) $\alpha: \mathbb{Z} \to \mathbb{Z}$, where $\alpha(m) = m - 4$;
(b) $\alpha: \mathbb{Z} \to \mathbb{Z}$, where $\alpha(m) = 2m + 3$;
(c) $\alpha: \mathbb{Q} \to \mathbb{Q}$, where $\alpha(q) = 2q + 3$;
(d) $\alpha: \mathbb{R} \to \mathbb{R}$, where $\alpha(x) = x^2$;
(e) $\alpha: \mathbb{R} \to \mathbb{R}$, where $\alpha(x) = x^3$;
(f) $\alpha: A \to A$, where $\alpha(a) = \dfrac{2a-1}{a-2}$, and the set $A = \mathbb{Q} - \{2\}$.
(g) $\alpha: \mathbb{Z} \times \mathbb{Z} \to \mathbb{Z}$, where $\alpha(m, n) = m + n$.

2 Let $\alpha, \beta: \mathbb{R} \to \mathbb{R}$, where $\alpha(x) = 2x - 3$ and $\beta(x) = 3x + 1$. Find
(a) the composite mapping $\gamma = \beta \circ \alpha$;
(b) the inverse mapping γ^{-1};
(c) the inverse mappings α^{-1} and β^{-1};
(d) the composite mapping $\alpha^{-1} \circ \beta^{-1}$ and verify that $\gamma^{-1} = \alpha^{-1} \circ \beta^{-1}$.

3 Let $X = \{x \in \mathbb{R}: x \geq 0\}$. Use theorem 3.4.3 to prove that the mapping $\gamma: X \to X$ defined by $\gamma(x) = x^2$ is bijective.

4 Let a, b be fixed real numbers and define $\alpha: \mathbb{R} \to \mathbb{R}$ by $\alpha(x) = ax + b$. Prove that α is bijective if and only if $a \neq 0$ and give α^{-1} in this case.

5 Let M be the set of all 2×2 matrices over $\mathbb{R}$.

Let $A = \begin{pmatrix} 2 & 1 \\ 1 & 1 \end{pmatrix}$ and $B = \begin{pmatrix} 2 & 2 \\ 1 & 1 \end{pmatrix}$; define mappings α, β: $M \to M$ such that $\alpha(X) = AX$ and $\beta(X) = BX$, for every matrix $X \in M$. Let $X = \begin{pmatrix} x & y \\ z & w \end{pmatrix} \in M$.

(a) Find $\alpha(X)$ and hence, or otherwise, prove that α is bijective. Find a matrix C such that $\alpha^{-1}(X) = CX$, for every $X \in M$.

(b) Find $\beta(X)$ and hence construct an example of two matrices X_1, X_2 such that $\beta(X_1) = \beta(X_2)$, but $X_1 \neq X_2$. Deduce that β is not invertible.

3.5 INDUCED MAPPINGS AND RESTRICTIONS

Induced mappings

Let A, B be non-empty sets and α: $A \to B$ any mapping. Then α gives rise in a natural way to a mapping of the subsets of A into the subsets of B. We shall denote this mapping by $\bar{\alpha}$, for the present. Dom $\bar{\alpha}$ is the set of all subsets of A, i.e. the power set $\mathcal{P}(A)$, and Codom $\bar{\alpha}$ is $\mathcal{P}(B)$. Then we define $\bar{\alpha}$ by

$$\bar{\alpha}: \mathcal{P}(A) \to \mathcal{P}(B),$$

where
$$\bar{\alpha}(X) = \begin{cases} \{\alpha(a): a \in X\}, & \text{when } X \neq \varnothing; \\ \\ \varnothing, & \text{when } X = \varnothing. \end{cases}$$

The mapping $\bar{\alpha}$ is called the mapping *induced* by α.

Examples 3.5.1

1. Let α: $\mathbb{Z} \to \mathbb{Z}$, where $\alpha(n) = 2n + 1$.
   Let $X = \{1\}$ $Y = \{1, 2\}$, $Z = \{10, 12, 35\}$.
   Then $\bar{\alpha}(X) = \{3\}$; $\bar{\alpha}(Y) = \{3, 5\}$; $\bar{\alpha}(Z) = \{21, 25, 71\}$.

2. Let β: $\mathbb{R} \to \mathbb{R}$, where $\beta(x) = x^2$.
   Let $[a, b]$ denote the set $\{x \in \mathbb{R}: a \leq x \leq b\}$.
   Then $\bar{\beta}([0, 1]) = [0, 1]$; $\bar{\beta}([-1, 0]) = [0, 1]$; $\bar{\beta}([-1, 3]) = [0, 9]$.

If α: $A \to B$, the induced mapping $\bar{\alpha}$: $\mathcal{P}(A) \to \mathcal{P}(A)$ is frequently denoted simply by α. Thus if $X \subseteq A$, it is common practice to write $\alpha(X)$ to denote the set $\{\alpha(a): a \in X\}$ instead of $\bar{\alpha}\{X\}$, although this is an abuse of notation. Note in particular that in this notation Im $\alpha = \alpha(A)$.

Restriction of a mapping

This is another simple idea which is perhaps best illustrated by some examples.

Examples 3.5.2

Let $X = \{x \in \mathbb{R}: x \geq 0\}$; $\quad C = \{x \in \mathbb{R}: -1 \leq x \leq 1\}$;

$$D = \left\{x \in \mathbb{R}: -\frac{\pi}{2} \leq x \leq \frac{\pi}{2}\right\}.$$

1. Let $\alpha: \mathbb{R} \to \mathbb{R}$, where $\alpha(x) = \sin x$.
   $\beta: \mathbb{R} \to C$, where $\beta(x) = \sin x$.
   $\gamma: D \to C$, where $\gamma(x) = \sin x$.

   Now although the formula for finding $\alpha(x)$, $\beta(x)$ and $\gamma(x)$ is the same, we cannot say that any pair of the mappings α, β and γ are equal, because no pair has the same domain and the same codomain. The mapping β is called the *restriction* of α to C as codomain. Similarly, γ is called the *restriction* of β to D as domain and it is also the *restriction* of α to D as domain and C as codomain.

2. Let

   $$\alpha: \mathbb{R} \to \mathbb{R}, \quad \text{where } \alpha(x) = x^2; \qquad \beta: \mathbb{Z}^+ \to \mathbb{Z}^+, \quad \text{where } \beta(x) = x^2.$$

   The mapping β is called the *restriction* of α to $\mathbb{Z}^+$ as domain and codomain.

Given a non-bijective mapping $\alpha: A \to B$, it may be possible to find a subset $X \subseteq A$ on which α is injective. If $\alpha(X) = Y$, then the restriction of α to X as domain and Y as codomain is bijective and hence invertible. Thus even when a mapping is not itself invertible, it may have an invertible restriction.

Example 3.5.3

Let α, β, γ be the mappings defined in examples 3.5.2, 1, above.
Then $\alpha: \mathbb{R} \to \mathbb{R}$ is neither injective nor surjective.

But $\bar{\alpha}(\mathbb{R}) = C$. Hence β is a *surjective* restriction of α.

Further, the restriction of α to D as domain is *injective*, and $\bar{\alpha}(D) = C$. Hence γ is a *bijective* restriction of α to D as domain and C as codomain. Thus although α is not an invertible mapping, it has an invertible restriction $\gamma: C \to D$. The inverse mapping $\gamma^{-1}: D \to C$ is denoted in calculus by $\sin^{-1} x$ or $\arcsin x$ and used as the inverse sine function.

PROBLEMS 3.5a

1 Let $A = \{1, 2, 3, 4\}$ and $\alpha: A \to A$ be defined by

$$\alpha(1) = \alpha(2) = 3, \qquad \alpha(3) = 4, \qquad \alpha(4) = 2.$$

Let $\bar{\alpha}: \mathscr{P}(A) \to \mathscr{P}(A)$ be the mapping induced by α on the subsets of A.

(a) Find the images of each of the following subsets under $\bar{\alpha}$:

(i) A;   (ii) $\{1\}$;   (iii) $\{3, 4\}$;   (iv) $\{1, 2, 3\}$.

(b) Show that $\bar{\alpha}$ is neither injective nor surjective.

2 Sketch a graph of the mapping $\alpha: \mathbb{R} \to \mathbb{R}$ defined by $\alpha(x) = x^2 - 2x$. With the notation of example 3.5.1, 2, find

(a) $\bar{\alpha}([0, 1])$;   (b) $\bar{\alpha}([1, 2])$;   (c) $\bar{\alpha}([0, 2])$;   (d) $\bar{\alpha}([0, 3])$.

3 Let $\gamma: X \to Y$ and suppose that $S \subseteq T \subseteq X$. Prove that $\bar{\gamma}(S) \subseteq \bar{\gamma}(T)$.

4 Let $\gamma: X \to Y$ and suppose that $S, T \subseteq X$. Prove that $\bar{\gamma}(S \cup T) = \bar{\gamma}(S) \cup \bar{\gamma}(T)$.

5 Suggest a bijective restriction of the mapping $\gamma: \mathbb{R} \to \mathbb{R}$ where $\gamma(x) = \exp(x)$. Give the inverse mapping of this restriction.

6 Show that the mapping $\alpha: \mathbb{R} \to \mathbb{R}$ where $\alpha(x) = x^2 + x$ is neither injective nor surjective. Suggest a bijective restriction $\gamma: S \to T$, where $S, T \subseteq \mathbb{R}$ and $T = \mathrm{Im}\,\alpha$.

PROBLEMS 3.5b

1 Sketch a graph of the mapping $\alpha: \mathbb{R} \to \mathbb{R}$, where $\alpha(x) = 2x + 3$. Find (a) $\bar{\alpha}([-2, 0])$; (b) $\bar{\alpha}([0, 3])$; (c) $\bar{\alpha}([-2, 3])$.

2 Sketch a graph of the mapping $\beta: \mathbb{R} \to \mathbb{R}$, where $\beta(x) = 3x - x^2$. Find (a) $\bar{\beta}([0,1])$; (b) $\bar{\beta}([0, 3])$; (c) $\bar{\beta}([1, 4])$.

3 Let $\alpha: A \to B$ be a mapping and $\bar{\alpha}: \mathcal{P}(A) \to \mathcal{P}(B)$ be the mapping induced by α.

(a) Show that if $\bar{\alpha}$ is not surjective, then α is not surjective. (*Hint*: Let $Y \subseteq B$ be a subset with no pre-image under $\bar{\alpha}$. Deduce that at least one element of Y has no pre-image under α.)

(b) Show that if α is not surjective, then $\bar{\alpha}$ is not surjective.
Deduce that $\bar{\alpha}$ is surjective $\Leftrightarrow$ α is surjective.

4 Let $\alpha: A \to B$ be a mapping and $\bar{\alpha}: \mathcal{P}(A) \to \mathcal{P}(B)$ be the mapping induced by α.

(a) Show that if α is not injective, then $\bar{\alpha}$ is not injective.

(b) Show that if $\bar{\alpha}$ is not injective, then α is not injective.
(*Hint*: Suppose S, $T \subseteq A$ are distinct subsets such that $\bar{\alpha}(S) = \bar{\alpha}(T)$. Show that there are distinct elements $s \in S$ and $t \in T$ such that $\alpha(s) = \alpha(t)$.)
Deduce $\bar{\alpha}$ is injective $\Leftrightarrow$ α is injective.

5 Let $\gamma: X \to Y$ and suppose S, $T \subseteq X$. Show that $\bar{\gamma}(S \cap T) \subseteq \bar{\gamma}(S) \cap \bar{\gamma}(T)$. Show, further, that if γ is injective, then $\bar{\gamma}(S \cap T) = \bar{\gamma}(S) \cap \bar{\gamma}(T)$. Let $X = \{1, 2, 3\}$ and $\gamma: X \to X$ be defined by $\gamma(1) = \gamma(2) = 1$ and $\gamma(3) = 2$. Find subsets S, $T \subseteq X$ such that $\bar{\gamma}(S \cap T) \neq \bar{\gamma}(S) \cap \bar{\gamma}(T)$.

SOLUTIONS TO PROBLEMS IN THE (a) SETS

Problems 3.1a

1 (b) is not a mapping because 2 is related to more than one element of T;

(c) is not a mapping because 3 is not related to any element of T.

2 (a) The mappings are (ii), (vii) and (ix).
(i) Every x is related to many values of y. (iii) $x < 0$ not related to any value of y or $x > 0$ related to two values of y. (iv) $x = 0$ not related to any value of y. (v) $|x| > 1$ not related to any value of y or $|x| < 1$ related to two values of y. (vi) $|x| > 1$ not related to any y. (viii) $|x| > 1$ not related to any y or $|x| < 1$ is related to many values of y. (x) Every $x < -1$ not related to any value of y or $x > -1$ related to two values of y.

(b) (ii) Im $\alpha = \{x \in \mathbb{R}: x \geq 0\}$;   (vii) Im $\alpha = \{x \in \mathbb{R}: |x| \leq 1\}$;
(ix) Im $\alpha = \{x \in \mathbb{R}; x \geq -1\}$.

3 (a) $\{-3\}$;   (b) $\{-1, 1\}$;   (c) $\{x \in \mathbb{R}: x \leq 0\}$;

(d) $\left\{(2k+1) \dfrac{\pi}{2}: k \in \mathbb{Z}\right\}$.

4 (b) and (d) are mappings.   (a) $\alpha(1)$ does not exist;
(c) 1 is related to both 0 and 2.

5 (a) $\{2n+1: n \in \mathbb{Z}^+\} = \{3, 5, 7, 9, \ldots\}$;

(b) $\left\{\dfrac{a}{2}: a \in \mathbb{Z}\right\} = \{0, \pm\frac{1}{2}, \pm\frac{2}{2}, \pm\frac{3}{2}, \ldots\}$;

(c) $\mathbb{Q}$;   (d) $\{n \in \mathbb{Z}: n \geq 3\}$;   (e) $\mathbb{Z}$.

6 (a) $2(2x) + 3 = 4x+3$;   (b) $2(x-1) + 3 = 2x + 1$;

(c) $\dfrac{2}{x} + 3$;   (d) $2x^2 + 3$.

Problems 3.2a

1 Injections are: (a), (b), (c), (d), (e). Surjections are (b), (d). Bijections are (b) and (d).

2 (b) (i) Neither; (ii) bijective; (iii) surjective; (iv) injective.

3 (a) α_2 and α_3. (b) Im $\alpha_1 = \{1, 3\}$;   Im $\alpha_2 = \{1, 2, 3\} =$ Im α_3; Im $\alpha_4 = \{3\}$. (c) α_2 and α_3. (d) No, since S is finite (see problem 5 below).

4 Suppose $\alpha(x_1) = \alpha(x_2)$. Then

$$\frac{x_1 + 1}{2x_1 - 1} = \frac{x_2 + 1}{2x_2 - 1} \Rightarrow (x_1 + 1)(2x_2 - 1) = (2x_1 - 1)(x_2 + 1)$$

$$\Rightarrow 2x_2 - x_1 = 2x_1 - x_2 \Rightarrow x_1 = x_2. \text{ Hence } \alpha \text{ is } injective.$$

Let $y \in \mathbb{R}$ and suppose $\alpha(x) = y$. Then

$$\frac{x+1}{2x-1} = y \Rightarrow x + 1 = 2xy - y$$

$$\Rightarrow x(2y - 1) = 1 + y \Rightarrow x = \frac{y + 1}{2y - 1}.$$

Hence when $y = \frac{1}{2}$, there is no $x \in \mathbb{R}$ such that $\alpha(x) = y$. Thus α is *not surjective*.

5 Let $S = \{x_1, x_2, \ldots, x_n\}$. Then Im $\alpha = \{\alpha(x_1), \alpha(x_2), \ldots, \alpha(x_n)\}$. Hence $|\text{Im } \alpha| \leq n$, and $|\text{Im } \alpha| = n \Leftrightarrow \alpha(x_1), \alpha(x_2), \ldots, \alpha(x_n)$ are all distinct $\Leftrightarrow$ α is *injective*. But Im $\alpha \subseteq S$ and hence Im $\alpha = S \Leftrightarrow |\text{Im } \alpha| = n$. Thus α is *surjective* $\Leftrightarrow |\text{Im } \alpha| = n \Leftrightarrow \alpha$ is *injective*.
(Follow the argument through referring to the mappings of problem 3.)
Counter examples: (a) $\beta: \mathbb{Z} \to \mathbb{Z}$ where $\beta(a) = 2a$ is injective but not surjective; (b) $\beta: \mathbb{R} \to \mathbb{R}$ where $\beta(x) = x^3 - x$ is surjective but not injective.

Problems 3.3a

1	$(\beta \circ a)(x)$	$(\alpha \circ \beta)(x)$	$(\alpha \circ \alpha)(x)$	$(\beta \circ \beta)(x)$
(a)	$2 \sin x + 1$	$\sin (2x + 1)$	$\sin (\sin x)$	$4x + 3$
(b)	x	x	$(x + 4)/9$	$9x - 4$
(c)	$x^2 - 2$	$(x - 2)^2$	x^4	$x - 4$

2 Im $\alpha = \{x \in \mathbb{R}: x \geq 0\}$; Im $\beta = \{x \in \mathbb{R}: -1 \leq x \leq 1\}$;
Im $\gamma = \mathbb{R}$. Since Dom $\gamma = \{x \in \mathbb{R}: x > 0\}$, we cannot define $\gamma \circ \alpha$
or $\gamma \circ \beta$, as Im $\alpha \not\subseteq$ Dom γ and Im $\beta \not\subseteq$ Dom γ.
$(\alpha \circ \beta)(x) = |\sin x|$; $(\beta \circ \alpha)(x) = \sin |x|$;
$(\alpha \circ \gamma)(x) = |\ln x|$; $(\beta \circ \gamma)(x) = \sin (\ln x)$.

3 First $\iota \circ \alpha$ and $\alpha \circ \iota$ are both mappings of A into A.
Let $a \in A$. Then $(\iota \circ \alpha)(a) = \iota(\alpha(a)) = \alpha(a)$, by definition of ι.
Hence $\iota \circ \alpha = \alpha$.
Also, $(\alpha \circ \iota)(a) = \alpha(\iota(a)) = \alpha(a)$, by definition of ι.
Hence $\alpha \circ \iota = \alpha$.

4 Converse of result 3.3.1: Let $\alpha: A \to B$ and $\beta: B \to C$.
Then $\beta \circ \alpha: A \to C$ is bijective $\Rightarrow \alpha, \beta$ are (both) bijective. Note: This
result is *false*. The rest of the question shows you how to construct a
counter-example.

(a) $\alpha (a_1) = \alpha(a_2) \Rightarrow (a_1 + 3, 1) = (a_2 + 3, 1) \Rightarrow a_1 = a_2$ (by definition
of *ordered pairs*) $\Rightarrow \alpha$ is injective. But $(c, b) \notin$ Im α when $b \neq$
1 and hence α is *not surjective*.
(b) Let c be any integer. Then $(c, 0) \in \mathbb{Z} \times \mathbb{Z}$ and $\beta (c, 0) = c$. Hence
Im $\beta = \mathbb{Z}$ and β is surjective. But $\beta (2, 1) = \beta (0, 3)$ and hence
β is *not injective*.
(c) $\beta \circ \alpha: \mathbb{Z} \to \mathbb{Z}$ where $(\beta \circ \alpha)(a) = \beta(\alpha(a)) = \beta(a + 3, 1) =$
$a + 4$. It is easy to show that this is bijective. But α is not bijective,
from (a), and β is not bijective from (b). Hence $\beta \circ \alpha$ bijective $\not\Rightarrow$
both α and β are bijective. (In fact our counter-example shows
that neither α nor β need be bijective.) $\alpha \circ \beta: \mathbb{Z} \times \mathbb{Z} \to \mathbb{Z} \times \mathbb{Z}$
where $(\alpha \circ \beta)(a, b) = \alpha(a + b) = (a + b + 3, 1)$; $\alpha \circ \beta$ is *not*
bijective.

Problems 3.4a

1 (a) $\alpha^{-1}: \mathbb{Z} \to \mathbb{Z}$, where $\alpha^{-1}(n) = n + 4$.
(b) Not invertible, since α is not surjective.
(c) $\alpha^{-1}: \mathbb{Q} \to \mathbb{Q}$, where $\alpha^{-1}(s) = (s - 3)/2$.
(d) Not invertible, since α is not injective or surjective.
(e) $\alpha^{-1}: \mathbb{R} \to \mathbb{R}$, where $\alpha^{-1}(y) = \sqrt[3]{y}$.

(f) $\alpha^{-1}: A \to A$, where $\alpha^{-1}(b) = \dfrac{2b-1}{b-2}$.

(g) Not invertible, since α is not injective.

2 (a) $\gamma: \mathbb{R} \to \mathbb{R}$ where $\gamma(x) = 6x - 8$;   (b) $\gamma^{-1}(x) = (8+x)/6$;

(c) $\alpha^{-1}(x) = (x+3)/2; \beta^{-1}(x) = (x-1)/3$;

(d) $(\alpha^{-1} \circ \beta^{-1})(x) = \alpha^{-1}\left(\dfrac{x-1}{3}\right)$

$$= \frac{1}{2}\left(\frac{x-1}{3} + 3\right) = \frac{x+8}{6} = \gamma^{-1}(x).$$

3 Define $\beta: X \to X$ by $\beta(x) = \sqrt{x}$. (Note that $\sqrt{x}$ is always taken to mean the *positive* square root of x.)

Then $(\beta \circ \gamma)(x) = \beta(x^2) = x; (\gamma \circ \beta)(x) = \gamma(\sqrt{x}) = x$.

Thus $\beta \circ \gamma = \iota = \gamma \circ \beta$, where ι is the identity mapping on X.

Hence γ is invertible and so bijective.

4 (i) Suppose $a \neq 0$. Define $\beta: \mathbb{R} \to \mathbb{R}$ by $\beta(x) = (x-b)/a$.

Then $(\beta \circ \alpha)(x) = \beta(ax+b) = x; (\alpha \circ \beta)(x) = \alpha\left(\dfrac{x-b}{a}\right) = x$.

Thus $\beta \circ \alpha = \iota = \alpha \circ \beta$, where ι is identity mapping on $\mathbb{R}$.

Therefore $a \neq 0 \Rightarrow \alpha$ is invertible and hence bijective.

(ii) Now suppose $a = 0$. Then $\alpha(x) = b, \forall x \in \mathbb{R}$. Hence Im $\alpha = \{b\}$ and α is neither injective nor surjective.

Thus α is bijective $\Leftrightarrow a \neq 0$. When $a \neq 0$, $\alpha^{-1}(x) = (x-b)/a$.

5 (a) $\alpha(X) = \begin{pmatrix} 2 & 1 \\ 1 & 1 \end{pmatrix}\begin{pmatrix} x & y \\ z & w \end{pmatrix} = \begin{pmatrix} 2x+z & 2y+w \\ x+z & y+w \end{pmatrix}$.

Let $W = \begin{pmatrix} a & b \\ c & d \end{pmatrix}$ be any matrix in M. Then $\alpha(X) = W$

$\Rightarrow 2x+z = a, x+z = c, 2y+w = b, y+w = d$, giving the *unique* solution $x = a-c, z = 2c-a, y = b-d, w = 2d-b$.

Hence there is a unique matrix $X = \begin{pmatrix} a-c & b-d \\ 2c-a & 2d-b \end{pmatrix}$ such that $\alpha(X) = W$.

The existence of $X \Rightarrow \alpha$ is surjective; the uniqueness of $X \Rightarrow \alpha$ is injective. Hence α is bijective and invertible.

Now $\alpha(X) = W \Rightarrow X = \alpha^{-1}(W)$. But $X = \begin{pmatrix} 1 & -1 \\ -1 & 2 \end{pmatrix}\begin{pmatrix} a & b \\ c & d \end{pmatrix}$.

Thus $\alpha^{-1}(W) = CW$, where $C = \begin{pmatrix} 1 & -1 \\ -1 & 2 \end{pmatrix}$.

(b) $\beta(X) = \begin{pmatrix} 2 & 2 \\ 1 & 1 \end{pmatrix}\begin{pmatrix} x & y \\ z & w \end{pmatrix} = \begin{pmatrix} 2x+2z & 2y+2w \\ x+z & y+w \end{pmatrix}$. To construct the example, choose $x_1 + z_1 = x_2 + z_2$ and $y_1 + w_1 = y_2 + w_2$.

e.g. let $X_1 = \begin{pmatrix} 1 & 2 \\ 3 & 4 \end{pmatrix}$. Then $x_1 + z_1 = 4$, $y_1 + w_1 = 6$.

Let $X_2 = \begin{pmatrix} 2 & 1 \\ 2 & 5 \end{pmatrix}$. Then $\beta(X_1) = \beta(X_2) = \begin{pmatrix} 8 & 12 \\ 4 & 6 \end{pmatrix}$. Hence β is not injective and so not invertible.

Problems 3.5a

1 (a) (i) $\{2, 3, 4\}$; (ii) $\{3\}$; (iii) $\{2, 4\}$; (iv) $\{3, 4\}$.
  (b) Note that α is not injective since $\alpha(1) = \alpha(2) = 3$. Hence
  $\bar\alpha: \{1\} \to \{3\}$ and $\{2\} \to \{3\}$. Thus $\bar\alpha$ is not injective. Note that α
  is not surjective, since $1 \notin \operatorname{Im} \alpha$. Hence $\{1\} \notin \operatorname{Im} \bar\alpha$ and so $\bar\alpha$ is
  not surjective.
  (There are several alternative counter-examples to the injective
  property, e.g. $\bar\alpha(\{1, 3\}) = \bar\alpha(\{2, 3\})$, $\bar\alpha(\{1, 3, 4\}) =$
  $\bar\alpha(\{2, 3, 4\})$ etc., and to the surjective property: $A \notin \operatorname{Im} \bar\alpha$ and
  no subset of A containing 1 belongs to $\operatorname{Im} \bar\alpha$.)

2 See Fig. 3.12.

  (a) $\bar\alpha([0, 1]) = [-1, 0]$;
  (b) $\bar\alpha([1, 2]) = [-1, 0]$;
  (c) $\bar\alpha([0, 2]) = [-1, 0]$;
  (d) $\bar\alpha([0, 3]) = [-1, 3]$.

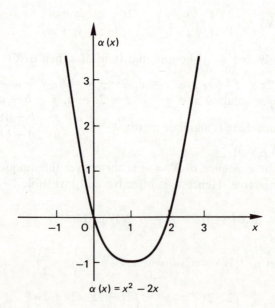

$\alpha(x) = x^2 - 2x$

Fig. 3.12

3 Let $y \in \bar{\gamma}(S) = \{\gamma(s): s \in S\}$. Thus $y = \gamma(s)$, for some $s \in S$. But $S \subseteq T$. Hence $s \in S \Rightarrow s \in T \Rightarrow \gamma(s) \in \bar{\gamma}(T)$. Thus $y \in \bar{\gamma}(S)$ $\Rightarrow y \in \bar{\gamma}(T)$. Hence $\bar{\gamma}(S) \subseteq \bar{\gamma}(T)$.

4 Let $y \in \bar{\gamma}(S \cup T) = \{\gamma(x): x \in S \cup T\}$. Thus $y = \gamma(x)$ for some $x \in S \cup T$. But $x \in S \cup T \Rightarrow x \in S$ or $x \in T \Rightarrow \gamma(x) \in \bar{\gamma}(S)$ or $\gamma(x) \in \bar{\gamma}(T) \Rightarrow \gamma(x) \in \bar{\gamma}(S) \cup \bar{\gamma}(T)$. Thus $y \in \bar{\gamma}(S \cup T)$ $\Rightarrow y \in \bar{\gamma}(S) \cup \bar{\gamma}(T)$. Hence (i) $\bar{\gamma}(S \cup T) \subseteq \bar{\gamma}(S) \cup \bar{\gamma}(T)$. Conversely, if $y \in \bar{\gamma}(S) \cup \bar{\gamma}(T)$, then $y \in \bar{\gamma}(S)$ or $y \in \bar{\gamma}(T)$.

If $y \in \bar{\gamma}(S)$, then $y = \gamma(x)$ for some $x \in S$; if $y \in \bar{\gamma}(T)$, then $y = \gamma(x)$ for some $x \in T$. But if $x \in S$ or $x \in T$, then $x \in S \cup T$. Hence $y \in \bar{\gamma}(S) \cup \bar{\gamma}(T) \Rightarrow y = \gamma(x)$, where $x \in S \cup T \Rightarrow y \in \bar{\gamma}(S \cup T)$.

Thus (ii) $\bar{\gamma}(S) \cup \bar{\gamma}(T) \subseteq \bar{\gamma}(S \cup T)$. Hence from (i) and (ii), $\bar{\gamma}(S \cup T)$ $= \bar{\gamma}(S) \cup \bar{\gamma}(T)$.

5 Since γ is injective there is no need to restrict the *domain* and we can make γ bijective by restricting the codomain to the set

$$\mathbb{R}^+ = \{x \in \mathbb{R}: x > 0\} = \text{Im } \gamma.$$

Then $\beta: \mathbb{R} \to \mathbb{R}^+$, where $\beta(x) = \exp(x)$ is a bijective restriction of γ; $\beta^{-1}: \mathbb{R}^+ \to \mathbb{R}$ is the mapping: $\beta^{-1}(x) = \ln x$.

6 $\alpha(x) = (x + \frac{1}{2})^2 - \frac{1}{4}$. Hence Im $\alpha = \{x \in \mathbb{R}; x \geq -\frac{1}{4}\}$, and α is not surjective. Since $\alpha(0) = \alpha(-1)$, α is not injective.

Let $S = \{x \in \mathbb{R}: x \geq -\frac{1}{2}\}$, $T = \text{Im } \alpha$. Then the restriction of α to S as domain, T as codomain is bijective.

4 THE INTEGERS

4.1 THE POSITIVE INTEGERS

Principle of induction

The basis of the proofs of many results involving the positive integers is the *principle of induction*. This depends upon the following two properties of $\mathbb{Z}^+$:

(i) $\mathbb{Z}^+$ has a least member (the number 1).
(ii) Given any number $n \in \mathbb{Z}^+$, there is a unique *next* (largest) number in $\mathbb{Z}^+$ (the number $n + 1$).

Of the other sets of numbers $\mathbb{Z}$, $\mathbb{Q}$, $\mathbb{R}$, $\mathbb{C}$, none possesses the first of these properties and only $\mathbb{Z}$ possesses the second. The principle of induction cannot therefore be used *by itself* to prove a result for 'any integer' or 'any rational' etc.

There are various formulations of the principle; the following is probably the most familiar.

Principle of induction (I)
Let $P(n)$ be a statement concerning the positive integer n. Suppose that we can prove that:

(i) $P(1)$ is true;
(ii) if $P(k)$ is true for some (particular) positive integer k then $P(k+1)$ is also true.

Then we can say that $P(n)$ is true for all positive integers.

Thus we first show that the result is true for the *first* number in $\mathbb{Z}^+$, and then show that if it is true for any particular number in $\mathbb{Z}^+$, it must also be true for the *next* number in $\mathbb{Z}^+$. This establishes the following 'chain' of true statements:

$P(1)$ is true, from (i);
since $P(1)$ is true, $P(2)$ is also true, from (ii);
since $P(2)$ is true, $P(3)$ is also true, from (ii);
since $P(3)$ is true, $P(4)$ is also true, from (ii),

and so on. Thus $P(n)$ is true for all $n \in \mathbb{Z}^+$.

Examples 4.1.1

1. Prove that $1^2 + 2^2 + 3^2 + \ldots + n^2 = n(n + 1)(2n + 1)/6$.

    Proof Let $P(n)$ be the statement we are required to prove.

    (i) Then $P(1)$ is the statement

$$1^2 = \tfrac{1}{6} \cdot 2 \cdot 3,$$

   which is clearly true.
    (ii) Suppose that $P(k)$ is true for some $k \in \mathbb{Z}^+$.
    Now $P(k+1)$ is the statement

$$1^2 + 2^2 + \ldots + (k+1)^2 = (k + 1)(k + 2)(2k + 3)/6.$$

But

$$
\begin{aligned}
1^2 + 2^2 + \ldots + (k + 1)^2 &= (1^2 + 2^2 + \ldots + k^2) + (k + 1)^2 \\
&= k(k + 1)(2k + 1)/6 + (k + 1)^2 \\
&\quad \text{(since } P(k) \text{ is true)} \\
&= (k + 1)(2k^2 + k + 6k + 6)/6 \\
&= (k + 1)(k + 2)(2k + 3)/6
\end{aligned}
$$

Thus if $P(k)$ is true, then $P(k + 1)$ is also true.
Hence by the principle of induction, $P(n)$ is true for all $n \in \mathbb{Z}^+$.

2. Prove that the total number of distinct subsets of a finite set of n elements is 2^n.

    We have established this result by two different methods in section 1.2. This third proof uses the principle of induction.

    Proof Let $P(n)$ be the statement: any set of n elements has exactly 2^n distinct subsets.

    (i) $P(1)$ is the statement: any set with 1 element has exactly 2 distinct subsets.

    If S is a set with just one element, then the only subsets of S are $\varnothing$ and S itself. Hence $P(1)$ is true.
    (ii) Suppose that for some $k \in \mathbb{Z}^+$, $P(k)$ is true. Then any set of k elements has exactly 2^k distinct subsets. Let S be a set of $k+1$ elements and let T be any subset of S containing just k of the elements of S; let x denote the one element of S not contained in T.

Then since $|T| = k$, T has exactly 2^k distinct subsets, since $P(k)$ is true. Denote these by $T_1, T_2, T_3, \ldots, T_{2^k}$.

Corresponding to each subset T_i, we define a set

$$X_i = \{x\} \cup T_i, i = 1, 2, \ldots, 2^k.$$

Clearly $T_i, X_i \subseteq S$, $i = 1, 2, \ldots, 2^k$ and the subsets $T_1, T_2, \ldots, T_{2^k}, X_1, X_2, \ldots, X_{2^k}$ are all distinct. Hence S contains at least $2 \cdot 2^k = 2^{k+1}$ distinct subsets. We must show this accounts for all the subsets of S.

Let $A \subseteq S$. Then if $x \notin A$, any element of A must be an element of T; hence $A \subseteq T$ and $A = T_i$, for some i, $1 \leq i \leq 2^k$; thus A has been accounted for. If $x \in A$, then we can write $A = \{x\} \cup B$, where $x \notin B$. Thus, as above $B = T_j$ for some j, $1 \leq j \leq 2^k$ and hence $A = X_j$, and again A has been accounted for. Hence S has exactly 2^{k+1} subsets and we have shown that if $P(k)$ is true, then $P(k+1)$ is also true. Thus from (i) and (ii), $P(n)$ is true for all n, by the principle of induction.

The statement of the principle of induction can easily be adapted to deal with the cases where the starting point for the 'chain of true statements', referred to above, is a number other than 1.

Principle of induction (I')

Let $P(n)$ be a statement concerning the integer n. Suppose we can prove:

  (i) $P(n_0)$ is true for some $n_0 \in \mathbb{Z}$;
 (ii) if $P(k)$ is true for some integer $k \geq n_0$, then $P(k + 1)$ is also true.

Then we can say that $P(n)$ is true for all integers $n \geq n_0$.

Example 4.1.2

Prove by induction that $2^n > n^2$ for all integers $n > 4$.

Proof Let $P(n)$ be the statement: $2^n > n^2$.
  (i) The least value of n for which we have to prove that $P(n)$ is true is $n = 5$.
      $P(5)$ is the statement: $2^5 > 5^2$, which is clearly true.
 (ii) Suppose $P(k)$ is true for some integer $k \geq 5$.
      Now $P(k + 1)$ is the statement: $2^{k+1} > (k + 1)^2$.
      But $2^{k+1} = 2 \cdot 2^k > 2k^2$, since $P(k)$ is true.
      However, $2k^2 = (k + 1)^2 + (k^2 - 2k - 1) > (k + 1)^2$, when $k \geq 5$.

Thus, when $k \geq 5$, $2^{k+1} > (k + 1)^2$.

Hence if $P(k)$ is true, where $k \geq 5$, then $P(k + 1)$ is also true and therefore $P(n)$ is true for all $n \geq 5$ by the principle of induction.

The generalised principle of induction

This form of induction can be used to prove all the examples we have so far given and a variety of other problems as well. It is therefore the most useful form of the principle to learn.

Suppose, as before, we are given a proposition $P(n)$ and required to show that $P(n)$ is true for all $n \geq 1$ (or for all $n \geq n_0$, for some fixed given integer n_0). We verify that $P(1)$ (or $P(n_0)$) is true, as before, and we could if necessary continue to verify the truth of $P(2)$, $P(3)$, . . ., for a finite number of cases. Hence either $P(n)$ is true for all $n \geq 1$ or there is a number k such that $P(n)$ is true for $1 \leq n < k$ but $P(k)$ is not true. The number k is called the *smallest counter-example*. If we assume such a number k exists and then show that $P(n)$ true for all n, where $1 \leq n < k$, implies $P(k)$ is also true, then we have a contradiction and hence $P(n)$ is true for all $n \geq 1$ (or for all $n \leq n_0$). This idea is stated formally below.

Principle of induction (II)

Let $P(n)$ be a statement concerning the positive integer n. Suppose we can prove:

  (i) that $P(1)$ is true;
  (ii) that if $P(n)$ is true, for all n such that $1 \leq n \leq k$, then $P(k)$ is also true.

Then $P(n)$ is true for all $n \in \mathbb{Z}^+$.

Examples 4.1.3

1. Every integer $n > 1$ can be expressed as the product of prime numbers.

   Note: A number $p \in \mathbb{Z}^+$ is prime if $p \geq 2$ and p has no factor $a \in \mathbb{Z}$ such that $2 \leq a < p$.

   Proof Let $P(n)$ be the statement: n can be expressed as the product of prime numbers.
     (i) Since 2 is itself prime, $P(2)$ is trivially true.
     (ii) Suppose that $P(n)$ is true for all n, where $2 \leq n < k$. Consider k. Either k is itself a prime, in which case $P(k)$ is trivially true, or k has a factor a such that $2 \leq a < k$ and hence we can write $k = ab$, where $2 \leq b < k$. Hence $P(a)$ and $P(b)$ are true and

we can express both a and b as the product of primes. Thus k can also be expressed as the product of primes and $P(k)$ is also true. Hence by the principle of induction (II), $P(n)$ is true for all $n \geq 2$.

Note: This result could not be proved directly from the induction principle (I), because the truth of $P(k)$ depends on the truth of both $P(a)$ and $P(b)$ where neither a nor b is $k - 1$.

2. The sequence $u_1, u_2, u_3, \ldots$ is defined inductively by the rule

$$u_n = 3u_{n-1} - 2u_{n-2}.$$

Given that $u_1 = 1$ and $u_2 = 2$, prove that $u_n = 2^n - 1$, $\forall\, n \in \mathbb{Z}^+$.

Proof   Note first that since the value of u_n depends on that of *two* previous terms, we must use the generalised principle of induction (II). Let $P(n)$ be the statement to be proved.

  (i) We cannot evaluate $u_3, u_4, \ldots$, unless both u_1 and u_2 are given and hence it is necessary to verify both $P(1)$ and $P(2)$. When $n = 1$, the formula gives $u_1 = 2 - 1 = 1$ and $u_2 = 2^2 - 1 = 3$, and hence $P(1)$ and $P(2)$ are both true.

 (ii) Suppose $P(n)$ is true where $1 \leq n < k$, for some integer $k > 2$. Then

$$\begin{aligned}
u_k &= 3u_{k-1} - 2u_{k-2} \\
&= 3(2^{k-1} - 1) - 2(2^{k-2} - 1) \quad \text{(since } P(k-1) \text{ and } P(k-2) \text{ are} \\
&\qquad\qquad\qquad\qquad\qquad\qquad\qquad \text{true, by assumption)} \\
&= 2 . 2^{k-1} - 1 \\
&= 2^k - 1.
\end{aligned}$$

Hence $P(n)$ true for $1 \leq n < k \Rightarrow P(k)$ is also true. Therefore $P(n)$ is true for all $n \in \mathbb{Z}^+$ by the principle of induction.

PROBLEMS 4.1a

1 Let $S_n = 1 + 3 + 5 + \ldots + (2n - 1)$. Calculate S_1, S_2, S_3 and S_4. Conjecture a simple formula for S_n and prove that it holds for all $n \in \mathbb{Z}^+$ by induction.

2 Prove that

$$\sum_1^n r^3 = [\tfrac{1}{2}n(n+1)]^2, \qquad \text{for all } n \in \mathbb{Z}^+.$$

3 Prove by induction that

$$\frac{1}{1.3} + \frac{1}{3.5} + \frac{1}{5.7} + \ldots + \frac{1}{(2n-1)(2n+1)} = \frac{n}{2n+1}, \qquad \text{for all } n \in \mathbb{Z}^+.$$

4 Prove by induction that $n! > 2^n$, for all integers $n > 3$.

5 Use the principle of induction to prove that for any real number $x > 0$ and any integer $n \geq 2$

$$(1 + x)^n > 1 + nx.$$

Deduce that $\left(1 + \dfrac{1}{n}\right)^n > 2$, for all $n \geq 2$.

6 Prove by induction that the integer $u_n = 4^n + 2$ is a multiple of 3, for every $n \in \mathbb{Z}$.

7 Let the matrix $A = \begin{pmatrix} 1 & a \\ 0 & 1 \end{pmatrix}$, where $a \in \mathbb{R}$, $a \neq 0$. Find A^2 and A^3 and conjecture a formula for A^n, $n \in \mathbb{Z}^+$. Prove your conjecture by the principle of induction.

8 The sequence $u_1, u_2, u_3, \ldots$, is defined inductively by the rule

$$u_n = 3u_{n-1} + 18u_{n-2}, \qquad n \geq 3.$$

Given that $u_1 = 6$ and $u_2 = 9$, prove by induction that u_n is a multiple of 3^n for all $n \in \mathbb{Z}^+$.

PROBLEMS 4.1b

1 Prove by induction that

$$\sum_1^n r = \tfrac{1}{2}n(n + 1), \qquad \text{for all } n \in \mathbb{Z}^+.$$

2 Let $S_n = \dfrac{1}{1.2} + \dfrac{1}{2.3} + \dfrac{1}{3.4} + \ldots + \dfrac{1}{n(n + 1)}$. Calculate S_1, S_2 and S_3. Conjecture a simple formula for S_n and prove that it holds for all $n \in \mathbb{Z}^+$ by induction.

3 Prove by induction that $2^n > n + 1$, for all integers $n > 1$.

4 Use the principle of induction to prove that for any real number $x \neq 1$, and for any $n \in \mathbb{Z}^+$,

$$1 + x + x^2 + \ldots + x^{n-1} = \dfrac{x^n - 1}{x - 1}.$$

5 Prove by induction that the integer $u_n = 4^n + 6n - 1$ is divisible by 9, for every $n \in \mathbb{Z}^+$.

6 Prove by induction that the integer $u_n = 7^n - 4^n$ is divisible by 3, for every $n \in \mathbb{Z}^+$.

7 The complex number $z = r(\cos \theta + i \sin \theta)$, where $r, \theta \in \mathbb{R}$. Prove by induction that $z^n = r^n(\cos n\theta + i \sin n\theta)$, for all $n \in \mathbb{Z}^+$.

8 The *Fibonacci sequence* is the sequence of integers $u_0, u_1, u_2, \ldots$, where $u_0 = 0$, $u_1 = 1$ and u_n is defined inductively by the formula

$$u_n = u_{n-1} + u_{n-2}, \qquad n \geq 2.$$

Thus each term in the sequence (after the second) is found by summing the previous two terms.

(a) Verify that the first six terms in the Fibonacci sequence are 0, 1, 1, 2, 3, 5. Find the next six terms.
(b) Prove by induction that

$$u_0 + u_2 + u_4 + \ldots + u_{2m} = u_{2m+1} - 1, \qquad m \geq 0.$$

(c) Prove by induction that

$$u_1 + u_3 + u_5 + \ldots + u_{2m-1} = u_{2m}, \qquad m \geq 1.$$

(d) Prove that u_n and u_{n+1} have no common factor >1, when $n \geq 1$. (*Hint*: Show that if h is a factor of both u_{k+1} and u_k, then h is also a factor of u_{k-1}, for any $k \geq 2$. Use a 'downwards' induction to show that if h is a factor of u_{n+1} and u_n, then h is also a factor of u_1.)
(e) Prove that $(u_{n+1})^2 - u_n u_{n+2} = (-1)^n$, for all $n \geq 0$.

4.2 MULTIPLES AND DIVISORS

Although we have already defined the terms *multiple* and *divisor* in Chapter 2, we repeat the definitions here for the sake of completeness and introduce a new symbol.

Multiples and divisors

Let $a, b \in \mathbb{Z}$. If there is an integer $q \in \mathbb{Z}$ such that $a = bq$ then we say that a is a *multiple* of b and that b is a *divisor* or *factor* of a. Symbolically this is written $b|a$.

Trivial factorisation

Clearly every integer a can be expressed as $a = a \times 1$ or as $a = (-a)(-1)$. This is called a *trivial factorisation*.

Units

We see that 1 and -1 occupy a special position in the study of factorisation of the integers as (a) they are factors of every integer $a \in \mathbb{Z}$ and (b) they are the *only* factors of 1 and -1, themselves. For these reasons, 1 and -1 are known as the *units* in $\mathbb{Z}$.

Example 4.2.1

We will illustrate the definition of *divisor* by asking

(a) Which integers are divisors of 0?
(b) Which integers are divisible by 0?

(a) From the definition, $a \in \mathbb{Z}$ is a *divisor* of 0 if we can find $b \in \mathbb{Z}$ such that

$$ab = 0.$$

But $b = 0$ gives $ab = 0$, $\forall \, a \in \mathbb{Z}$.
Hence *every* integer is a divisor of 0; symbolically, $a|0$, $\forall \, a \in \mathbb{Z}$.
Equivalently, we can say that 0 is a *multiple* of *every* integer.

(b) From the definition, $a \in \mathbb{Z}$ is *divisible* by 0 if we can find $c \in \mathbb{Z}$ such that

$$c0 = a.$$

But $c0 = 0$, $\forall \, c \in \mathbb{Z}$. Hence the *only* integer divisible by 0 is 0 itself; symbolically, $0|a \Rightarrow a = 0$. Equivalently, we can say that the *only* multiple of 0 is 0 itself.

Prime numbers

If $p \in \mathbb{Z}$, $p \neq \pm 1$, is such that p has only a trivial factorisation, then p is said to be a *prime* number.

Composite integers

If $a \in \mathbb{Z}$ has a non-trivial factorisation, then a is called a *composite integer*.

Thus every integer is either *composite*, *prime* or a *unit*.

The division theorem

The basic result on division of integers is known as the *division theorem*. The proof we give uses the well-ordering principle, which is stated below. This principle can be proved to be logically equivalent to the principle of induction: that is, each can be deduced from the other.

The well-ordering principle
Let S be a non-empty subset of $\mathbb{Z}$ such that for all $n \in S$, $n \geq n_0$, where n_0 is some fixed integer. Then S has a *least* member.

THEOREM 4.2.1 (Division theorem)
Given any integers a, b, with $b > 0$, we can find unique integers q, r, where $0 \leq r < b$, such that $a = bq + r$. Also, $b|a$ if and only if $r = 0$.

Proof　　Rewriting the equation as $r = a - bq$, we see that we are looking for an integer q such that $0 \leq a - bq < b$. Hence we consider the set $S = \{a - bx : x \in \mathbb{Z}\}$ and try to show that S contains an integer r such that $0 \leq r < b$.

First we show that S contains at least one non-negative integer.

(i) Suppose $a \geq 0$. Then since $x = 0$ gives $a \in S$, S contains a non-negative integer.

(ii) Suppose $a < 0$. Then putting $x = a$ gives $a - ab \in S$. But $a - ab = (-a)(b - 1) \geq 0$, since $-a > 0$ and $b \geq 1$. Hence S contains a non-negative integer in this case too.

Now let $T = \{t \in S : t \geq 0\}$. Then we have shown above that T is non-empty and hence T satisfies the conditions of the well-ordering principle. Thus T contains a least member, t_0, say. Then $t_0 \geq 0$, by the definition T. Also, since $t_0 \in S$, $t_0 = a - bx_0$, for some $x_0 \in \mathbb{Z}$.

We now show that $t_0 < b$. Suppose if possible $t_0 \geq b$.
Consider the integer $t = a - b(x_0 + 1) \in S$.
Now $t = a - bx_0 - b = t_0 - b \Rightarrow 0 \leq t < t_0$.

Hence $t \in T$. But $t < t_0$ contradicts the choice of t_0 as the least element of T. Thus $t_0 \geq b$ is impossible and we conclude $t_0 < b$.

Thus taking $r = t_0$ and $q = x_0$, we have shown there exists at least one pair of integers r, q satisfying $a - bq = r$ and $0 \leq r < b$.

Finally we must show that r and q are unique. Suppose we can find another pair of integers r_1, q_1 satisfying $a - bq = r_1$ and $0 \leq r_1 < b$.
Then $|r_1 - r| = |(a - bq_1) - (a - bq)|$
$\qquad\qquad = b|q - q_1|$.
Now $q_1 \neq q \Rightarrow |q - q_1| \geq 1 \Rightarrow |r - r_1| \geq b$.

But $|r - r_1| \geq b$ is impossible since $0 \leq r < b$ and $0 \leq r_1 < b$. Thus we must have $q_1 = q$, which forces $r_1 = r$, also. This proves the uniqueness of q and r.

Finally it is clear that $r = 0 \Rightarrow a = bq$. But $a = bq$ means $b|a$. This concludes the proof of the theorem. ∎

Examples 4.2.2

1. We illustrate the proof of the division theorem by taking a numerical example. Suppose $a = -13$, $b = 5$.

   Let $S = \{a - bx: x \in \mathbb{Z}\}$
   $$= \{-13 - 5x: x \in \mathbb{Z}\}$$
   $$= \{\ldots, -18, -13, -8, -3, 2, 7, 12, 17, \ldots\}.$$

   By inspection S contains some positive integers.

   Let $T = \{t \in S: t \geq 0\}$
   $$= \{2, 7, 12, 17, \ldots\}.$$

   The least positive member of T is 2.

   We can express 2 as: $2 = -13 - 5(-3)$.

   Thus $-13 = 5(-3) + 2$, giving $q = -3$, $r = 2$, satisfying $-13 = 5q + r$ and $0 \leq r < 5$.

2. Let $n \in \mathbb{Z}$. Prove that one of the integers $n - 1, n, n + 1$ is divisible by 3.

   Proof By the division theorem, we can find q, r, with $0 \leq r < 3$. such that

   $$n = 3q + r.$$

   There are thus three possibilities to consider: $r = 0, 1$ or 2.

      (i) $r = 0$ gives $n = 3q$ and hence $3|n$;
      (ii) $r = 1$ gives $n = 3q + 1$ and hence $n - 1 = 3q$ and $3|n - 1$;
      (iii) $r = 2$ gives $n = 3q + 2$ and hence $n + 1 = 3(q+1)$ and $3|n + 1$.

   Thus in any case, one of the integers $n - 1, n$ and $n + 1$ is divisible by 3.

3. Let $a, b, d \in \mathbb{Z}$. Prove that if $d|a$ and $d|b$, then (a) $d|a+b$ and (b) $d|a - b$.

   Proof $d|a \Rightarrow$ we can find $s \in \mathbb{Z}$ such that $a = ds$.
           $d|b$ we can find $t \in \mathbb{Z}$ such that $b = dt$.
   Hence (a) $a + b = ds + dt = d(s + t) \Rightarrow d|a + b$.
          (b) $a - b = ds - dt = d(s - t) \Rightarrow d|a - b$.

PROBLEMS 4.2a

1 Find the integers q, r in the division theorem when $b=4$ and a is
(a) 19; (b) -19; (c) 1; (d) -1; (e) 0.

2 Let a, b, q, $r \in \mathbb{Z}$ and suppose that $a = bq + r$. Suppose $d \in \mathbb{Z}$, $d|a$ and $d|b$. Use the method of example 4.2.2, 3, to prove that $d|r$.

3 Let a, b, $d \in \mathbb{Z}$ and suppose $d|a$ and $d|b$. Use the method of example 4.2.2, 3, to prove that $d|ma + nb$, $\forall$ m, $n \in \mathbb{Z}$.

4 Let a, b, $c \in \mathbb{Z}$ and suppose $a|b$ and $b|c$. Prove that $a|c$.

5 Let a, $b \in \mathbb{Z}$ and suppose $a|b$ and $b|a$. Show that $a = \pm b$.

6 Taking $b = 2$ in the division theorem, show that every integer a can be written either as $2q$ or as $2q+1$, where $q \in \mathbb{Z}$.

7 Let n, $b \in \mathbb{Z}$ and consider the b successive integers n, $n + 1$, $n + 2$, $\ldots$, $n + b - 1$. Apply the division theorem to $a = n + b - 1$ and b to show that $bq = n + b - 1 - r$ for some $q, r \in \mathbb{Z}$, where $0 \leq r < b$. Deduce that just one of any b successive integers is divisible by b. (*Hint*: See example 4.2.2, 2, where this result is proved for the particular case when $b = 3$.)

4.3 GREATEST COMMON DIVISOR

Common divisor

Let a, b, $c \in \mathbb{Z}$. If $c|a$ and $c|b$, then c is called a *common divisor* (or *common factor*) of a and b.

Given $a \in \mathbb{Z}$, $a \neq 0$ let $D(a)$ denote the set of divisors of a. Then if $t \in D(a)$, we have $-|a| \leq t \leq |a|$ and hence $D(a)$ is a finite set. The set of *common divisors* of a, $b \in \mathbb{Z}$ is a subset of $D(a)$ and clearly will also be a finite set. Hence it has a greatest member.

Greatest common divisor

Let a, $b \in \mathbb{Z}$, $a \neq 0$, $b \neq 0$. Let g be a positive integer such that

(a) $g|a$ and $g|b$;
(b) if c is a common divisor of a and b, then $g \geq c$.

Then g is called the *greatest common divisor* (or g.c.d.) of a and b. It is usually denoted by (a, b), or, to avoid confusion with other notations, by $\gcd(a, b)$.

Note: If $a \neq 0$, then the largest positive integer d such that $d|a$ and $d|0$ is a. Hence we define $\gcd(a, 0) = a$. We do not define $\gcd(0, 0)$.

Examples 4.3.1

By inspection: $\gcd(88, 64) = 8$; $\gcd(-18, 6) = 6$;
$\qquad\qquad\quad \gcd(20, -7) = 1$; $\gcd(8, 0) = 8$.

It is easy to show the following and the proofs are left to the reader.

RESULT 4.3.1   Let $a, b \in \mathbb{Z}$, $b \neq 0$. Then

(a) $\gcd(a, b) = \gcd(b, a)$;
(b) $\gcd(a, b) = \gcd(-a, b) = \gcd(a, -b)$
$\qquad\qquad\qquad\qquad = \gcd(-a, -b)$;
(c) $\gcd(1, b) = 1$;
(d) if p is prime and $p \nmid b$, then $\gcd(p, b) = 1$
(e) if $b|a$, then $\gcd(a, b) = |b|$.

Coprime integers

Let a, b be non-zero integers. If $\gcd(a, b) = 1$, then a and b are said to be *coprime* or *relatively prime*.

Euclid's algorithm

If a and b are small (positive or negative) integers, then it is often possible to find the sets $D(a)$ and $D(b)$ of their divisors by trial and error and hence determine $\gcd(a, b)$ by inspection. However, this method can be tedious for large integers. A method for finding $\gcd(a, b)$ which does not depend upon knowing the factors of a and b is known as *Euclid's algorithm*. It dates back to about 300 BC.

From result 4.3.1, $\gcd(a, b) = \gcd(|a|, |b|)$ and hence there is no loss of generality in assuming that both a and b are positive.

THEOREM 4.3.1 (Euclid) Let a, $b \in \mathbb{Z}^+$. Then there exist unique integers q_i, r_i, $i = 1, 2, \ldots, n + 1$, where $b > r_1 > r_2 > \ldots > r_n > r_{n+1} =$ such that

$$a = bq_1 + r_1$$
$$b = r_1q_2 + r_2$$
$$r_1 = r_2q_3 + r_3$$
$$\ldots$$
$$\ldots$$
$$\ldots$$
$$r_{n-2} = r_{n-1}g_n + \boxed{r_n} \quad \leftarrow \quad \gcd(a, b)$$
$$r_{n-1} = r_nq_{n+1} + r_{n+1}.$$

Then $\gcd(a, b) = r_n$, the *last non-zero remainder*.

Proof   We apply the division theorem first to the pair a, b, giving unique integers q_1, r_1, such that

$$a = bq_1 + r_1, \quad \text{where } 0 \le r_1 < b. \qquad \ldots (1)$$

If $r_1 \ne 0$, we apply the division theorem again to the pair b, r_1, giving unique integers q_2, r_2 such that

$$b = r_1q_2 + r_2, \quad \text{where } 0 \le r_2 < r_1 < b. \qquad \ldots (2)$$

If $r_2 \ne 0$, we apply the division theorem again to the pair r_1, r_2, giving unique integers q_3, r_3 such that

$$r_1 = r_2q_3 + r_3, \quad \text{where } 0 \le r_3 < r_2 < r_1 < b. \qquad \ldots (3)$$

Continuing in this way, the remainders form a strictly decreasing sequence. But no remainder is less than 0. Hence after a finite number of steps, we must reach a remainder r_{n+1}, say, such that $r_{n+1} = 0$.

Then the last steps are:

$$r_{n-1} = r_{n-2}q_{n-1} + r_{n-1}, \quad \text{where } 0 < r_{n-1} < r_{n-2} < \ldots < b, \qquad \ldots (4)$$

$$r_{n-2} = r_{n-1}q_n + r_n, \quad \text{where } 0 < r_n < r_{n-1} < \ldots < b, \qquad \ldots (5)$$

$$r_{n-1} = r_nq_{n+1} + 0. \qquad \ldots (6)$$

Now from equation (6), $r_n | r_{n-1}$.

Hence in equation (5), $r_n | r_{n-1} \Rightarrow r_n | r_{n-2}$.

Similarly, in equation (4), $r_n | r_{n-1}$ and $r_n | r_{n-2} \Rightarrow r_n | r_{n-3}$.

Continuing in this way, we work our way back to equation (2), where we have

$$r_n | r_2 \quad \text{and } r_n | r_1 \Rightarrow r_n | b,$$

and in equation (1),

$$r_n|r_1 \text{ and } r_n|b \Rightarrow r_n|a.$$

Thus r_n is a common divisor of a and b.

To show that r_n is the greatest common divisor, we shall show that if c is any common divisor of a and b, then $c|r_n$ and hence $c \le r_n$.

First, we rearrange equations (1) to (5), and write:

$$r_1 = a - bq_1; \qquad \ldots (1)$$

$$r_2 = b - r_1q_2; \qquad \ldots (2)$$

$$r_3 = r_1 - r_2q_3; \qquad \ldots (3)$$

$$\begin{array}{ccc} \cdot & \cdot & \cdot \\ \cdot & \cdot & \cdot \\ \cdot & \cdot & \cdot \end{array}$$

$$r_{n-1} = r_{n-3} - r_{n-2}q_{n-1}; \qquad \ldots (4)$$

$$r_n = r_{n-2} - r_{n-1}q_n. \qquad \ldots (5)$$

Let $c|a$ and $c|b$. Then from (1), $c|r_1$. Hence in (2), $c|b$ and $c|r_1 \Rightarrow c|r_2$, and in (3), $c|r_1$ and $c|r_2 \Rightarrow c|r_3$.

Continuing, we see that c is a divisor of every remainder, until finally, in (5), $c|r_{n-2}$ and $c|r_{n-1} \Rightarrow c|r_n$.

But $c|r_n \Rightarrow |c| \le r_n$ and hence $r_n = \gcd(a, b)$. ∎

Examples 4.3.2

1. We illustrate the algorithm by using it to calculate $\gcd(88, 64)$, which we know by inspection is 8.

   First, note that since $\gcd(a, b) = \gcd(b, a)$, we may always choose a to be the larger of the integers.

   Let $a = 88$, $b = 64$. Then by the division theorem:

   (a) $88 = 64(1) + 24$
   (b) $64 = 24(2) + 16$
   (c) $24 = 16(1) + 8$ ←————— last non-zero remainder
   (d) $16 = 8(2)$.

   Hence, by Euclid's algorithm, $\gcd(88, 64) = 8$.

   In this example we have written the quotients q_i in brackets; the *quotients* each appear in one equation only, whereas the *remainders* appear in *successive* equations, moving at each appearance one place to the left. It is a good idea, therefore, to use a notation which distinguishes in some way between those integers which will not be used again and those that will.

2. Prove that 1234 and 491 are coprime.

Solution   Apply Euclid's algorithm with $a = 1234$ and $b = 491$.

$$1234 = 491(2) + 252$$
$$491 = 252(1) + 239$$
$$252 = 239(1) + 13$$
$$239 = 13(18) + 5$$
$$13 = 5(2) + 3$$
$$5 = 3(1) + 2$$
$$3 = 2(1) + 1 \leftarrow \text{last non-zero remainder}$$

Hence $\gcd(1234, 491) = 1$ and these integers are *coprime*.

In the last part of the proof of Euclid's algorithm, we established that *every* common divisor of a and b is a divisor of their g.c.d. This is such a useful fact that we now repeat it separately:

RESULT 4.3.2   Let c be any common divisor of the non-zero integers a, b. Then $c|\gcd(a, b)$.

Another useful result is the following:

RESULT 4.3.3   Let a, b be non-zero integers and suppose $\gcd(a, b) = g$. Let $a = ga_1$ and $b = gb_1$. Then $\gcd(a_1, b_1) = 1$.

Proof   Suppose $\gcd(a_1, b_1) = d$.
Then $a_1 = da_2$ and $b_1 = db_2$, for some $a_2, b_2 \in \mathbb{Z}$.
Hence $a = gda_2$ and $b = gdb_2$.
Thus $gd|a$ and $gd|b$. Hence $gd|g$, by result 4.3.2.
But $gd|g \Rightarrow d = \pm 1 \Rightarrow \gcd(a_1, b_1) = 1$, as required. ∎

This establishes the fairly obvious fact that if we divide a pair of non-zero integers by their *greatest* common divisor then the resulting pair of quotients will be relatively prime.

Linear combinations

Let a, b be non-zero integers. Then any integer that can be written in the form

$$ax + by, \quad \text{where } x, y \in \mathbb{Z}$$

is called a *linear combination* of a and b.

Examples 4.3.3

1. Find all the integers that can be written as linear combinations of 10 and 25.

   Solution   Let S be the set of all linear combinations of 10 and 25.
   Then $S = \{10x + 25y: x, y \in \mathbb{Z}\}$.
   Let $m \in S$. Then $m = 10x + 25y$, for some particular values of x and y. By inspection, $\gcd(10, 25) = 5$.
   Writing $m = 5(2x + 5y)$, we see that $5|m$, whatever the values of x and y. Hence $5|m$, $\forall\, m \in S$.
   Thus every member of S is a multiple of 5.
   But does S contain *every* multiple of 5?
   To answer this in the affirmative, we show that 5 itself is a member of S.
   By inspection,

   $$5 = 10(3) + 25(-1). \qquad \cdots \text{(1)}$$

   Hence $5 \in S$.
   (Notice that there are many other possible choices for x and y.)
   Any multiple of 5 can be written as $5k$, for some $k \in \mathbb{Z}$.

   But from (1), $5k = k\{10(3) + 25(-1)\}$
                    $= 10(3k) + 25(-k)$.

   Hence $5k \in S, \forall\, k \in \mathbb{Z}$.

   Thus $S = \{5k: k \in \mathbb{Z}\}$.

2. Prove that every integer can be written as a linear combination of 5 and 13.

   Solution   We first think of a way of writing 1 as a linear combination of 5 and 13. There are many ways; one such is the following:

   $$1 = 13(2) + 5(-5).$$

   Hence, given any integer $k \in \mathbb{Z}$,

   $$k = k\{13(2) + 5(-5)\}$$
   $$= 13(2k) + 5(-5k).$$

   Thus k can be written as a linear combination of 5 and 13.

The next result generalises the first part of the solution of example 4.3.3, 1.

RESULT 4.3.4 Let a, b be non-zero integers and let $g = $ gcd(a, b). Then every linear combination of a and b is a multiple of g. That is, $g|ax + by$, $\forall x, y \in \mathbb{Z}$.

Proof   Since g is a common divisor of a and b, we can write $a = ga_1$ and $b = gb_1$, for some $a_1, b_1 \in \mathbb{Z}$.
Then $ax + by = g(a_1x + b_1y)$.
Hence $g|ax + by$, $\forall x, y \in \mathbb{Z}$. ∎

In order to generalise the second part of the solution of example 4.3.3, 1, we must prove that gcd(a, b) can be written as a linear combination of a and b. This is another of the fundamental results of number theory. There are two, at least, different methods of proving this result. The proof we give below gives an actual algorithm for determining a pair of integers x, y such that gcd$(a, b) = ax + by$.

THEOREM 4.3.5 Let a, b be non-zero integers and let $g = $ gcd(a, b). Then we can find integers x, y such that

$$g = ax + by.$$

Proof   By Euclid's algorithm, there exist integers $q_i, r_i, i = 1, 2, \ldots, n$, such that

$$r_1 = a - bq_1 \qquad \ldots (1)$$

$$r_2 = b - r_1q_2 \qquad \ldots (2)$$

$$\vdots \qquad \vdots \qquad \vdots$$

$$r_{n-2} = r_{n-4} - r_{n-3}q_{n-2} \qquad \ldots (3)$$

$$r_{n-1} = r_{n-3} - r_{n-2}q_{n-1} \qquad \ldots (4)$$

$$g = r_n = r_{n-2} - r_{n-1}q_n. \qquad \ldots (5)$$

Using equation (4), we can substitute in (5) for r_{n-1}, giving

$$g = r_{n-2} - (r_{n-3} - r_{n-2}q_{n-1})q_n$$
$$= r_{n-2}(1 + q_{n-1}q_n) - r_{n-3}q_n.$$

Using (3), we substitute for r_{n-2} and hence obtain an expression for g as a linear combination of r_{n-3} and r_{n-4}. Continuing in this way we eventually substitute for r_2 from (2), obtaining an expression for g as a linear combination of b and r_1, and substituting for r_1 from (1), we finally express g as *a* linear combination of a and b. ■

Examples 4.3.4

1. Given that gcd(88, 64) = 8, write 8 as a linear combination of 88 and 64.

Solution   By Euclid's algorithm, we have (see example 4.3.2,1)

$$24 = 88 - 64(1) \qquad \qquad \cdots (1)$$
$$16 = 64 - 24(2) \qquad \qquad \cdots (2)$$
$$8 = 24 - 16(1). \qquad \qquad \cdots (3)$$

In this example $a = 88$, $b = 64$, $r_1 = 24$, $r_2 = 16$, $r_3 = 8$. Equation (3) expresses 8 as a linear combination of 16 and 24.

Using (2), we can substitute for 16 in (3) and hence express 8 as a linear combination of 24 and 64.

$$8 = 24 - \{64 - 24(2)\}\,(1)$$
$$= 24(3) - 64(1).$$

Next, using (1), we can substitute in this equation for 24 and hence write 8 as a linear combination of 64 and 88 as required.

$$8 = \{88 - 64(1)\}\,(3) - 64(1)$$
$$= 88(3) - 64(4).$$

Thus $8 = 88x + 64y$ where $x = 3$ and $y = -4$.

2. Find integers s, t such that

$$1 = 1234s + 491t.$$

Solution   We showed that gcd(1234, 491) = 1, by Euclid's algorithm in example 4.3.2, 2. Turning each of those equations round we have

$$252 = 1234 - 491(2) \qquad \qquad \cdots (1)$$
$$239 = 491 - 252(1) \qquad \qquad \cdots (2)$$
$$13 = 252 - 239(1) \qquad \qquad \cdots (3)$$
$$5 = 239 - 13(18) \qquad \qquad \cdots (4)$$
$$3 = 13 - 5(2) \qquad \qquad \cdots (5)$$
$$2 = 5 - 3(1) \qquad \qquad \cdots (6)$$
$$1 = 3 - 2(1). \qquad \qquad \cdots (7)$$

Using (6), we first substitute for 2 in (7), giving

$$
\begin{aligned}
1 &= 3 - \{5 - 3(1)\}(1) \\
&= 3(2) - 5(1) \\
&= \{13 - 5(2)\}(2) - 5(1), \quad \text{using (5),} \\
&= 13(2) - 5(5) \\
&= 13(2) - \{239 - 13(18)\}(5), \quad \text{using (4),} \\
&= 13(92) - 239(5) \\
&= \{252 - 239(1)\}(92) - 239(5), \quad \text{using (3),} \\
&= 252(92) - 239(97) \\
&= 252(92) - \{491 - 252(1)\}(97), \quad \text{using (2),} \\
&= 252(189) - 491(97) \\
&= \{1234 - 491(2)\}(189) - 491(97) \\
&= 1234(189) - 491(475).
\end{aligned}
$$

Hence $1 = 1234s + 491t$, where $s = 189$ and $t = -475$.

THEOREM 4.3.6   Let a, b be non-zero integers and let $g = \gcd(a, b)$. Then the set of all linear combinations of a and b is equal to the set of all multiples of g.

Proof   Let $S = \{ax + by: x, y \in \mathbb{Z}\}$ and $T = \{gk: k \in \mathbb{Z}\}$. We want to prove that $S = T$.

(i) By result 4.3.4, $m \in S \Rightarrow m \in T$, and hence $S \subseteq T$.

(ii) But by theorem 4.3.5, $g \in S$. Hence $g = ax_0 + by_0$, for some $x_0, y_0 \in \mathbb{Z}$.

Now $t \in T \Rightarrow t = gk$,   for some $k \in \mathbb{Z}$.

$$
\begin{aligned}
&\Rightarrow t = (ax_0 + by_0)k. \\
&\quad\ = a(x_0 k) + b(y_0 k) \\
&\Rightarrow t \in S.
\end{aligned}
$$

Thus $T \subseteq S$.

Hence from (i) and (ii), $S = T$. ∎

With the notation of theorem 4.3.6, it is clear that g is the least positive integer in T and hence, since $S = T$, g is also the least positive integer in S. Thus we have

---

COROLLARY 4.3.7   The smallest positive integer that can be written as a linear combination of two non-zero integers a, b is $\gcd(a, b)$.

PROBLEMS 4.3a

1 Show by the method of example 4.3.3, 2 that any integer k can be written as a linear combination of (a) 5 and 7; (b) 8 and 13.

2 Use Euclid's algorithm to show that $\gcd(38, 16) = 2$. Hence find a formula for expressing any even integer $2n$ as a linear combination of 38 and 16.

3 Use Euclid's algorithm to show $\gcd(1547, 3059) = 7$. Hence find integers x, y such that
$$7 = 1547x + 3059y.$$

4 Use Euclid's algorithm to find integers s, t such that
$$1 = 1001s + 30t.$$
Hence prove that $\{1001s + 30t: s, t \in \mathbb{Z}\} = \mathbb{Z}$.

5 Use Euclid's algorithm to find a pair of integers x, y such that
$$19x + 15y = 1.$$
Hence show how to write 6 as a linear combination of
    (a) 19 and 15;   (b) 114 and 90;   (c) 38 and 30.
Is it possible to write 7 as a linear combination of any of these three pairs of numbers? Justify your answer.

6 Use Euclid's algorithm to find $\gcd(51, 39)$.
Hence find integers x, y such that
$$39x + 51y = 15.$$
Show that it is not possible to find integers x, y such that
$$39x + 51y = 5.$$

7 Let a, b, d be non-zero integers and suppose $a = a_1 d$ and $b = b_1 d$ where $a_1, b_1 \in \mathbb{Z}$. Prove that if $\gcd(a_1, b_1) = 1$, then $d = \gcd(a, b)$. (This is the *converse* of result 4.3.3.)

(*Hint*: Start by saying:

$$\gcd(a_1, b_1) = 1 \Rightarrow 1 = a_1 x + b_1 y, \quad \text{for some } x, y \in \mathbb{Z}.$$

Then multiply through this identity by d. Hence show $\gcd(a, b) | d$.)

PROBLEMS 4.3b

1 Show that the set of integers which can be written as a linear combination of 54 and 30 is equal to the set of multiples of a certain integer m, and find the value of m.

    Hence give a formula for writing 48 as a linear combination of 54 and 30.

2 Use Euclid's algorithm to show that gcd(5643, 4851) = 99.
Find integers s, t such that
$$99 = 5643s + 4851t.$$

3 Use Euclid's algorithm to show that 498 and 179 are coprime.

4 Use Euclid's algorithm to show that 1764 and 605 are coprime and hence find integers x, y such that
$$1764x + 605y = 1.$$

5 Use Euclid's algorithm to find integers s, t such that
$$1 = 17s + 14t.$$
Hence write 8 as a linear combination of 17 and 14.

6 Find integers x, y such that
$$63x + 49y = 70.$$
Is it possible to find integers s, t such that
$$63s + 49t = 60?$$
Either find the integers s, t or prove they cannot exist.

7 Let a, b, m be non-zero integers such that a and m are coprime.
Prove that if $m|ab$, then $m|b$.
(*Hint*: Start by saying:
$$\gcd(a, m) = 1 \Rightarrow 1 = ax + my, \quad \text{for some } x, y \in \mathbb{Z}.$$
Then multiply through this identity by b.)

8 Let a, b, m be non-zero integers and suppose that $\gcd(a, b) = 1$.
Prove that if $a|m$ and $b|m$, then $ab|m$. Give an example to show that if $\gcd(a, b) > 1$, then it is possible to have $a|m$ and $b|m$, but $ab \nmid m$.
(*Hint*: Start by saying:
$$\gcd(a, b) = 1 \Rightarrow 1 = as + bt, \text{ for some } s, t \in \mathbb{Z}.$$
Then multiply through by m.)

4.4 LINEAR CONGRUENCES

In section 2.4, we introduced *congruence relations* on the integers. A *linear* congruence is a congruence relation of the form

$$ax \equiv b \pmod{m},$$

where a, b, x, $m \in \mathbb{Z}$, $a \neq 0$, $m > 0$. The congruence is called *linear* because the variable x occurs only to the power 1, as in a linear equation.

In this section, we shall show that linear congruences can be manipulated to some extent like linear equations (see result 4.4.1, below), but that, unlike a linear equation, a linear congruence does not always have a solution. We shall use the results and methods of the last section to prove a simple test for deciding whether a given linear congruence has a solution and a method for finding the general solution in this case.

RESULT 4.4.1   Let $a \equiv b \pmod{m}$, $c \equiv d \pmod{m}$, $t \in \mathbb{Z}$. Then

(a) $a + t \equiv b + t \pmod{m}$;

(b) $at \equiv bt \pmod{m}$;

(c) $a \pm c \equiv b \pm d \pmod{m}$;

(d) $ac \equiv bd \pmod{m}$.

Proof $a \equiv b \pmod{m} \Rightarrow a = b + hm$, for some $h \in \mathbb{Z}$.
Similarly, $c \equiv d \pmod{m} \Rightarrow c = d + km$, for some $k \in \mathbb{Z}$.
Hence

(a) $a + t = (b + t) + hm \Rightarrow a + t \equiv b + t \pmod{m}$;

(b) $at = (b + hm)t = bt + (ht)m \Rightarrow at \equiv bt \pmod{m}$;

(c) $a + c = b + d + (h + k)m \Rightarrow a + c \equiv b + d \pmod{m}$;

$a - c = b - d + (h - k)m \Rightarrow a - c \equiv b - d \pmod{m}$;

(d) $ac = (b + hm)(d + km) = bd + (bk + hd + hkm)m$
$\Rightarrow ac \equiv bd \pmod{m}$. ■

RESULT 4.4.2   Let $\gcd(a, m) = g$. Then the linear congruence

$$ax \equiv b \pmod{m}$$

has a solution *only if* $g|b$.

Proof   Let $a = ga_1$, and $m = gm_1$.
Suppose there exists a solution $x = x_0$ of the linear congruence
$ax \equiv b \pmod{m}$. Then $ax_0 \equiv b \pmod{m}$.
Hence $ax_0 = b + mt$, for some $t \in \mathbb{Z}$, and so b
$$= ax_0 - mt$$
$$= g(a_1 x_0 - m_1 t)$$
$$\Rightarrow g|b.$$
Hence a solution exists $\Rightarrow g|b$.
Or, equivalently, $g \nmid b \Rightarrow no$ solution exists. ■

RESULT 4.4.3   Let $\gcd(a, m) = g$ and suppose $g|b$. Then the
linear congruence

$$ax \equiv b \pmod{m}$$

has a solution.

Proof Let $b = gb_1$. Then by result 4.3.5, since $g = \gcd(a, m)$ we can find integers s, t such that

$$g = as + mt$$
$$\Rightarrow gb_1 = asb_1 + mtb_1$$
$$\Rightarrow b = a(sb_1) + m(tb_1)$$
$$\Rightarrow a(sb_1) = b + m(-tb_1).$$

Hence $x = sb_1$ is one solution of the linear congruence

$$ax \equiv b \ (\text{mod } m)$$

Thus $g|b \Rightarrow a$ solution exists. ∎

Clearly if $g = 1$, then $g|b$, whatever the integer b. Hence we have the following useful corollary to result 4.4.3.

COROLLARY 4.4.4 If $\gcd(a, m) = 1$, the linear congruence

$$ax \equiv b \ (\text{mod } m)$$

always has a solution.

We now have a very simple test for deciding whether the linear congruence $ax \equiv b \ (\text{mod } m)$ has a solution or not: we find $g = \gcd(a, m)$, by inspection or by Euclid's algorithm, and check whether $g|b$. Combining results 4.4.2 and 4.4.3 gives

---

THEOREM 4.4.5 Let $g = \gcd(a, m)$. Then the linear congruence

$$ax \equiv b \ (\text{mod } m)$$

has a solution if and only if $g|b$.

In the following theorem we suppose that $g|b$ and prove a formula for the general solution.

THEOREM 4.4.6 Let $\gcd(a, m) = g$, $a = ga_1$, $m = gm_1$. Suppose $g|b$ and that $x = x_0$ is a solution of the linear congruence $ax \equiv b \ (\text{mod } m)$. Then the general solution is

$$x = x_0 + km_1, \quad k \in \mathbb{Z}.$$

Proof Since $g|b$, we know that a solution x_0 exists.
But $x = x_0$ is a solution of $ax \equiv b \pmod{m}$.

$$\Rightarrow ax_0 = b + y_0 m, \quad \text{for some } y_0 \in \mathbb{Z}. \qquad \dots (1)$$

(i) We first verify that any integer of the form $x_0 + km_1$ is also a solution.

$$\begin{aligned} a(x_0 + km_1) &= ax_0 + akm_1 \\ &= (b + y_0 m) + akm_1, \quad \text{from (1)}. \end{aligned}$$

Now
$$akm_1 = ka_1 gm_1 = ka_1 m,$$

so
$$a(x_0 + km_1) = b + y_0 m + ka_1 m$$
$$= b + m(y_0 + ka_1).$$

Hence
$$a(x_0 + km_1) \equiv b \pmod{m}.$$

Thus
$$x = x_0 + km_1 \quad \text{is a solution}, \quad k \in \mathbb{Z}.$$

(ii) We must next ask: can every solution be written in the form $x_0 + km_1$ or are there other solutions as well?

In other words, does the set $\{x_0 + km_1 : k \in \mathbb{Z}\}$ contain all the solutions of $ax \equiv b \pmod{m}$? To find out, let us suppose that $x = x_0 + c$ is a solution. Then if we can show that $m_1|c$, we will have proved that every solution is indeed of the form $x_0 + km_1$, for some $k \in \mathbb{Z}$.

Now $x = x_0 + c$ is a solution of $ax \equiv b \pmod{m}$

$\Rightarrow \quad a(x_0 + c) = b + ym$, for some $y \in \mathbb{Z}$.

$\Rightarrow (b + y_0 m) + ac = b + ym$, from (1)

$\Rightarrow ac = m(y - y_0)$

$\Rightarrow a_1 c = m_1 (y - y_0)$, dividing both sides by g.

$\Rightarrow m_1 | a_1 c$.

But $\gcd(a_1, m_1) = 1$, by result 4.3.3. This means that m_1 and a_1 have no common factors other than ± 1. Hence $m_1 | a_1 c \Rightarrow m_1 c$ (see problem 4.3b, 7). But $m_1 | c \Rightarrow c = km_1$, for some $k \in \mathbb{Z}$, and hence every solution is of the form $x_0 + km_1$, $k \in \mathbb{Z}$. ∎

We see from result 4.4.2 and theorem 4.4.6 that unlike a linear equation, a linear congruence never has a *unique* solution. It has either no solution or an infinity of solutions. If we can find any one solution x_0, by inspection or calculation, then any integer in the congruence class $\bar{x}_0 \pmod{m_1}$ is also a solution. We sometimes need the *least non-negative* solution and this is clearly the least non-negative integer in $\bar{x}_0 \pmod{m_1}$. Thus we have:

COROLLARY 4.4.7   Let $\gcd(a, m) = g$ and $m = m_1 g$. Then if $g|b$, the linear congruence

$$ax \equiv b \ (\text{mod } m)$$

has a (unique) solution in the range $0 \leq x < m_1$.

Examples 4.4.1

Find, where possible, the general solution and least non-negative solution of the following linear congruences:

(a) $5 \equiv 4 \ (\text{mod } 9)$.

Solution $\gcd(5, 9) = 1$ and hence a solution exists.

By corollary 4.4.7, there is a (unique) solution in the range $0 \leq x < 9$.

As the numbers are small, it is probably quickest to find this solution by trial and error (see examples 2.4.3).

Least non-negative solution: $x = 8$. General solution $x = 8 + 9k$, $k \in \mathbb{Z}$.

(b) $5x \equiv 4 \ (\text{mod } 10)$.

Solution $\gcd(5, 10) = 5$. But $5 \nmid 4$. Hence no solution exists.

(c) $24x \equiv 18 \ (\text{mod } 39)$.

Solution   By inspection, $\gcd(24, 39) = 3$. As $3|18$, a solution exists. Hence there exist integers x, y such that

$$24x = 18 + 39y. \qquad \ldots (1)$$

But every pair of integers x, y satisfying (1) also satisfy:

$$8x = 6 + 13y$$

(dividing through (1) by $g = 3$), giving

$$6 = 8x - 13y. \qquad \ldots (2)$$

Now $\gcd(8, 13) = 1$ and hence we can write every integer as a linear combination of 8 and 13, using the methods of section 4.3. Applying Euclid's algorithm to 13 and 8 gives:

$$13 = 8(1) + 5$$
$$8 = 5(1) + 3$$
$$5 = 3(1) + 2$$
$$3 = 2(1) + 1.$$

Hence $1 = 3 - 2(1)$, and retracing our steps in the usual way gives

$$1 = 8(5) - 13(3)$$
$$6 = 8(30) - 13(18).$$

Comparing this identity with equation (2) shows that $x = 30$ is a solution of (2) and hence of (1).

Now $m_1 = m/g = 39/3 = 13$.

Hence the general solution is $x = 30 \pmod{13}$.

Since $30 = 13(2) + 4$, the least non-negative solution is $x = 4$.

(d) $224x \equiv 154 \pmod{385}$.

Solution We use Euclid's algorithm to determine $\gcd(385, 224)$.

$$385 = 224(1) + 161$$
$$224 = 161(1) + 63$$
$$161 = 63(2) + 35$$
$$63 = 35(1) + 28$$
$$35 = 28(1) + 7$$
$$28 = 7(4).$$

Hence $\gcd(385, 224) = 7$. Since $7 | 154$, a solution exists.

Hence we can find x, y such that

$$224x = 154 + 385y. \qquad \ldots (1)$$

We could divide through this equation by $g = 7$ (as in the previous example) but the coefficients of x and y would still be large and we would probably have to apply Euclid's algorithm to these new coefficients in order to find x and y. Hence in this case it is quicker to use the working above to write 7 as a linear combination of 385 and 224. We have

$$7 = 35 - 28(1)$$

and continuing in the usual way gives

$$7 = 385(7) - 224(12).$$

Now $154 = 7.22$, so multiplying through by 22 gives

$$154 = 385(154) - 224(264)$$
$$224(-264) = 154 + 385(-154). \qquad \ldots (2)$$

Comparing equations (1) and (2) shows that $x = -264$ is a solution.

Now $m_1 = m/g = 385/7 = 55$.

Hence the general solution is $x = 55k - 264$, and the least non-negative solution is $x = 11$.

PROBLEMS 4.4a

1 Show that
(a) $n \equiv 7 \pmod{12} \Rightarrow n \equiv 3 \pmod{4}$;
(b) $n \equiv 10 \pmod{18} \Rightarrow n \equiv 1 \pmod{3}$;
(c) $n \equiv 13 \pmod{24} \Rightarrow n \equiv 1 \pmod{6}$.

2 Using result 4.4.1 (d), prove by induction that if $a \equiv b \pmod{m}$, then $a^n \equiv b^n \pmod{m}$, for all $n \in Z^+$. Hence prove that

(a) $10^n \equiv 1 \pmod 9$, for all $n \in Z^+$;

(b) $10^n \equiv 1 \pmod 3$, for all $n \in Z^+$;

(c) $10^n \equiv \begin{cases} 1 \pmod{11} \text{ if } n \text{ is even,} \\ 1 \pmod{11} \text{ if } n \text{ is odd,} \end{cases} \quad n \in Z;$

(d) every positive integer power of 5 is just 1 greater than a number exactly divisible by 4;

(e) every *even* power of 9 ends in the digit 1.

3 Show that each of the following linear congruences $ax \equiv b \pmod m$ has the same set of solutions as a linear congruence of the form $a_1 x \equiv b_1 \pmod{m_1}$, where $\gcd(a_1, m_1) = 1$. Hence find the general solution of each.

(a) $8x \equiv 12 \pmod{20}$;   (b) $9x \equiv 3 \pmod{15}$;

(c) $10x \equiv 6 \pmod{24}$;   (d) $10x \equiv 15 \pmod{25}$;

(e) $14x \equiv 35 \pmod{63}$;   (f) $18x \equiv 6 \pmod{42}$.

4 Decide whether each of the following linear congruences has a solution. If so, find the general solution and least positive solution.

(a) $15x \equiv 6 \pmod{21}$;   (b) $15x \equiv 6 \pmod{40}$;

(c) $15x \equiv 6 \pmod{36}$;   (d) $15x \equiv 6 \pmod 7$.

5 Use Euclid's algorithm to find integers x, y such that $11x = 1 + 63y$. Hence find the general solution of

(a) $11x \equiv 1 \pmod{63}$;   (b) $11x \equiv 5 \pmod{63}$.

6 Find integers x, y such that $13x = 1 + 22y$. Hence find the general solution and the least positive integer x satisfying:

(a) $13x \equiv 1 \pmod{22}$;   (b) $13x \equiv 3 \pmod{22}$;

(c) $13x \equiv 10 \pmod{22}$.

7 Find the least positive integer x satisfying

$$84x \equiv 48 \pmod{192}.$$

PROBLEMS 4.4b

1 Decide whether each of the following congruences has a solution. If so, find the general solution and the least non-negative solution.

(a) $12x \equiv 9 \pmod{15}$;   (b) $12x \equiv 9 \pmod{16}$;

(c) $12x \equiv 9 \pmod{17}$;   (d) $12x \equiv 9 \pmod{18}$;

(e) $12x \equiv 9 \pmod{25}$;   (f) $12x \equiv 9 \pmod{27}$.

2 Find integers x, y such that $17x = 1 + 24y$. Hence find the least positive integer x satisfying each of the following congruences:

(a) $17x \equiv 1 \pmod{24}$;   (b) $17x \equiv 5 \pmod{24}$;
(c) $17x \equiv 10 \pmod{24}$.

3 Find the general solution and the least positive integer x satisfying

$$135x \equiv 63 \pmod{207}.$$

4 Let m be a positive integer with $n + 1$ digits,
say $m = a_n a_{n-1} \ldots a_1 a_0$.
Then we can write

$$m = a_n 10^n + a_{n-1} 10^{n-1} + \ldots + a_1 10 + a_0, \quad 0 \le a_i \le 9,$$
$$i = 1, 2, \ldots, n, a_n \neq 0.$$

Hence show that

(a) $m \equiv a_0 \pmod{2}$
(b) $m \equiv 10a_1 + a_0 \pmod{4}$
(c) $m \equiv 100a_2 + 10a_1 + a_0 \pmod{8}$.

Deduce the following tests for divisibility:

(a) an integer is divisible by 2 if its final digit is divisible by 2;
(b) an integer is divisible by 4 if the number formed by the final two digits is divisible by 4;
(c) an integer is divisible by 8 if the number formed by the last three digits is divisible by 8.

5 Using result 4.4.1 and problem 4.4.a, 2(a), prove that

(a) $a_k 10^k \equiv a_k \pmod{9}$,   (b) $\displaystyle\sum_{k=0}^{n} a_k 10^k \equiv \sum_{k=0}^{n} a_k \pmod{9}$,

where $a_k \in \mathbb{Z}, k = 0, 1, \ldots, n$.

Hence, using the notation of problem 4 above, show that an integer is divisible by 9 if the sum of its digits is divisible by 9.

Using problem 4.4a, 2(a) and (c), prove in a similar way that an integer is divisible (a) by 3 if the sum of its digits is divisible by 3 and (b) by 11 if the difference between the sums $a_0 + a_2 + a_4 + \ldots$ and $a_1 + a_3 + a_5 + \ldots$ is divisible by 11.

6 Prove by induction that

$$5^n \equiv 5 \pmod{10}, \quad \forall \, n \in \mathbb{Z}^+.$$

Hence show that if the final digit of an integer m is 5, then the final digit of m^n is also 5, for all $n \in \mathbb{Z}^+$.

Find three other digits for which a similar result is true, proving the result in each case.

4.5 PRIMES AND UNIQUE FACTORISATION

In section 4.1 we proved by induction that every positive integer n ≥ 2 is prime or can be expressed as the product of primes (example 4.1.3, 1). Primes are in this sense the building blocks of the integers. In this section we shall establish a few elementary properties of prime numbers, including the *fundamental theorem of arithmetic* which states that not only can we factor $n \geq 2$ into primes, but this factorisation is unique. In order to prove this we need the following result.

RESULT 4.5.1   Let p be a prime and a, b non-zero integers. Then if $p|ab$, $p|a$ or $p|b$.

Proof   Suppose that $p{\nmid}a$. Then since p is prime, $\gcd(a, p) = 1$. Hence we can find integers s, t such that

$$1 = as + pt.$$
Then
$$b = bas + bpt.$$

Now $p|ab \Rightarrow ab = pr$, for some $r \in \mathbb{Z}$.
Hence $b = p(rs + bt)$,   so that $p|b$. ∎

We can generalise result 4.5.1 to a product of more than two factors. The proof uses result 4.5.1 and induction, and is left as an exercise (see problems 4.5.a).

RESULT 4.5.2   Let p be a prime and $a_1, a_2, \ldots, a_n$ be non-zero integers. Then $p|a_1 a_2 \ldots a_n \Rightarrow p|a_i$, for some i, $1 \leq i \leq n$.

THEOREM 4.5.3   (Fundamental theorem of arithmetic) Let $n \in \mathbb{N}$, $n \geq 2$. Then

(a) n can be expressed as the product of one or more primes;
(b) this expression is unique, except for the order in which the factors are listed.

Proof (a) We have already proved this by induction in example 4.1.3, 2.

(b) We will prove this, also, by the generalised principle of induction. Let $P(n)$ be the statement: the expression for n as a product of primes is unique (except for the order in which they are listed). Then $P(2)$ is true.

Suppose that $P(n)$ is true for all integers n such that $2 \leq n < k$. Suppose that $k = p_1 p_2 \ldots p_r$ and $k = q_1 q_2 \ldots q_s$ are two prime factorisations of k. Then $p_1 | k$ and hence $p_1 | q_1 q_2 \ldots q_s$. But p_1 is a prime and so by result 4.5.2, $p_1 | q_1 q_2 \ldots q_s \Rightarrow p_1 | q_i$ for some i, $1 \leq i \leq s$. But q_i is also a prime, so that $p_1 | q_i \Rightarrow p_1 = q_i$.

Now consider $k_1 = k/p_1 = k/q_i$. Since $2 \leq k_1 < k$, $P(k_1)$ is true and so the list of primes $p_2, p_3, \ldots, p_r$ is the same as the list q_1, $q_2, \ldots, q_{i-1}, q_{i+1}, \ldots, q_s$ (except possibly for order). Thus $r = s$ and the two factorisations of k contain exactly the same primes. Thus $P(n)$ true for $2 \leq n < k \Rightarrow P(k)$ is also true and hence $P(n)$ is true for all $n \geq 2$. ∎

In view of this theorem, it is important to ask whether there are an infinite number of primes at our disposal as building blocks of the integers. The following theorem answers the question. The proof given here is attributed to Euclid (*c.* 300 BC).

THEOREM 4.5.4   There are infinitely many primes.

Proof   Suppose that there is only a finite number of primes, say p_1, $p_2, \ldots, p_N$. Consider the number $n = p_1 p_2 \ldots p_N + 1$. If n is prime then it is a new prime since $n > p_i$, $i = 1, 2, \ldots, N$. If n is not prime then it has a prime factor p such that $2 \leq p < n$, by the fundamental theorem. If p is not a new prime then it must be one of the primes in the list $p_1, p_2, \ldots, p_N$, say $p = p_k$, $1 \leq k \leq N$. Then $p_k | n$ and $p_k | n-1$ and hence $p_k | n - (n-1) = 1$. This contradiction shows that p is a new prime. Hence our list of primes can never be complete. ∎

Primes have fascinated mathematicians throughout history. The proofs of many of the results on primes are complicated and require a knowledge of analysis. We conclude this section with a result which has a very simple proof based on result 4.5.1.

THEOREM 4.5.5   Let $p > 0$ be a prime. Then $\sqrt{p}$ is irrational.

Proof This is another example of a proof by contradiction. Suppose that $\sqrt{p}$ is rational. Then we can find natural numbers a, b such that $\sqrt{p} = a/b$, where we may assume that a/b is a fraction 'in its lowest terms', so that a and b are coprime.

But $\sqrt{p} = a/b \Rightarrow p = a^2/b^2$
$$\Rightarrow pb^2 = a^2$$
$$\Rightarrow p|a^2$$
$$\Rightarrow p|a, \quad \text{by result 4.5.1.}$$

Hence we can write $a = pa_1$, where $a_1 \in \mathbb{Z}$. Thus

$$pb^2 = a^2 \Rightarrow pb^2 = p^2a_1^2$$
$$\Rightarrow b^2 = pa_1^2$$
$$\Rightarrow p|b^2$$
$$\Rightarrow p|b, \quad \text{by result 4.5.1.}$$

But this implies that p is a common factor of a and b, contradicting the assumption that a and b are coprime. Hence $\sqrt{p}$ is not rational when p is prime. ∎

One further idea, which we will need later in the study of permutations, is defined below. It is closely related to the g.c.d. (see problem 4.5a, 3).

Least common multiple

Let a, b be non-zero integers. Let m be *a* positive integer such that

(a) $a|m$ and $b|m$;
(b) if $r \in \mathbb{Z}^+$, $a|r$ and $b|r$, then $m \leqslant r$.

We denote the *least common multiple* of a and b by $\text{lcm}(a, b)$.

PROBLEMS 4.5a

1 Using result 4.5.1, prove result 4.5.2 by induction on the number n of factors in the product $a_1a_2 \ldots a_n$.
  Let p be a prime. Deduce that
  $$p|a^n \Rightarrow p|a, \quad \text{for all } n \in \mathbb{Z}^+.$$

2 Use the method of theorem 4.5.5 to show that $\sqrt{10}$ is irrational.

3 Let a, b be non-zero integers and let $g = \gcd(a, b)$. Prove that
  $$\text{lcm}(a, b) = ab/g.$$

SOLUTIONS TO PROBLEMS IN THE (a) SETS

Problems 4.1a

The following are skeleton solutions only. In particular, the straight-forward verification that $P(1)$ is true is left to the reader. Where appropriate, S_n denotes $\sum_1^n u_r$.

1 $S_1 = 1$; $S_2 = 4$; $S_3 = 9$; $S_4 = 16$. Let $P(n)$ be the statement: $S_n = n^2$.
 (i) Check $P(1)$ true.
 (ii) Assume $P(k)$ true for some $k \in \mathbb{Z}^+$.
   Then $S_{k+1} = S_k + (2k + 1) = k^2 + (2k + 1) = (k + 1)^2$.
   Hence $P(k)$ true $\Rightarrow P(k + 1)$ true.

2 Let $P(n)$ be the statement: $S_n = [\frac{1}{2}n(n + 1)]^2 = \frac{1}{4}n^2(n + 1)^2$.
 (i) Check $P(1)$ true.
 (ii) Assume $P(k)$ true for some $k \in \mathbb{Z}^+$.
   Then $S_{k+1} = S_k + (k + 1)^3 = \frac{1}{4}k^2(k + 1)^2 + (k + 1)^3$
   $\qquad\qquad = \frac{1}{4}(k + 1)^2 (k^2 + 4(k + 1)) = \frac{1}{4}(k + 1)^2(k + 2)^2$.
   Hence $P(k)$ true $\Rightarrow P(k + 1)$ true.

3 Let $P(n)$ be the statement: $S_n = \dfrac{n}{2n + 1}$.

 (i) Check $P(1)$ true.
 (ii) Suppose $P(k)$ true for some $k \in \mathbb{Z}^+$. Then

$$S_{k+1} = S_k + \frac{1}{(2(k + 1)-1)(2(k + 1)+1)}$$

$$= \frac{k}{2k + 1} + \frac{1}{(2k + 1)(2k + 3)}$$

$$= \frac{2k^2 + 3k + 1}{(2k + 1)(2k + 3)} = \frac{k + 1}{(2k + 3)}.$$

   Hence $P(k)$ true $\Rightarrow P(k + 1)$ true.

4 Let $P(n)$ be the statement: $n! > 2^n$.
 (i) Check $P(4)$ true.
 (ii) Suppose $P(k)$ true for some $k \geq 4$.
   Then $(k + 1)! = (k + 1)(k!) > (k + 1)2^k > 2.2^k$, since $k \geq 4$.
   Hence $P(k)$ true $\Rightarrow P(k + 1)$ true, $k \geq 4$.

5 Let $P(n)$ be the statement: $(1 + x)^n > 1 + nx$. Then
 (i) $(1 + x)^2 = 1 + 2x + x^2 > 1 + 2x$, since $x^2 > 0$. Hence $P(2)$ is true.
 (ii) Suppose $P(k)$ is true for some $k \geq 2$. Then
   $(1 + x)^{k+1} = (1 + x)(1 + x)^k > (1 + x)(1 + kx) =$
   $1 + (k + 1)x + kx^2 > 1 + (k + 1)x$, since $kx^2 > 0$. Hence $P(k)$
   true $\Rightarrow P(k + 1)$ true, $k \geq 2$.

Putting $x = \dfrac{1}{n}$, $n \geq 2$, gives $\left(1 + \dfrac{1}{n}\right)^n > 1 + \dfrac{n}{n} = 2$.

6 Let $P(n)$ be the statement: $u_n = 4^n + 2$ is divisible by 3.

(i) Check u_1 is divisible by 3.

(ii) Suppose $P(k)$ true for some $k \in \mathbb{N}$. Then $u_k = 4^k + 2 = 3m$, say.

Then $u_{k+1} = 4^{k+1} + 2 = 4.4^k + 2 = 4(3m - 2) + 2 = 12m - 6 = 3(4m - 2)$.

Hence u^k is divisible by $3 \Rightarrow u_{k+1}$ is divisible by 3.

7
$$A^2 = \begin{pmatrix} 1 & 2a \\ 0 & 1 \end{pmatrix}; \quad A^3 = \begin{pmatrix} 1 & 3a \\ 0 & 1 \end{pmatrix}.$$

Let $P(n)$ be the statement: $A^n = \begin{pmatrix} 1 & na \\ 0 & 1 \end{pmatrix}$. Then

(i) $P(1)$ is true.

(ii) Suppose $P(k)$ is true for some $k \in \mathbb{Z}^+$. Then

$$A^{k+1} = A \times A^k = \begin{pmatrix} 1 & a \\ 0 & 1 \end{pmatrix}\begin{pmatrix} 1 & ka \\ 0 & 1 \end{pmatrix} = \begin{pmatrix} 1 & (k+1)a \\ 0 & 1 \end{pmatrix}.$$

Hence $P(k)$ true $\Rightarrow P(k + 1)$ true.

8 Let $P(n)$ be the statement: u_n is a multiple of 3^n.

(i) Check *both* that u_1 is a multiple of 3 and u_2 is a multiple of 3^2.

(ii) Suppose that $P(n)$ is true for all $n < k$, for some $k \geq 3$.

Then we can write $u_{k-2} = 3^{k-2}s$ and $u_{k-1} = 3^{k-1}t$.

Hence $u_k = 3.3^{k-1}t + 18.3^{k-2}s = 3^k(t + 2s)$. Hence $P(n)$ true for all $n < k \Rightarrow P(k)$ true, $k \geq 3$. But $P(1)$ and $P(2)$ are true. Hence $P(n)$ true for all $n \in \mathbb{Z}^+$.

Problems 4.2a

1 (a) $q = 4$, $r = 3$; (b) $q = -5$, $r = 1$; (c) $q = 0$, $r = 1$; (d) $q = -1$, $r = 3$; (e) $q = r = 0$.

2 Let $a = a_1d$, $b = b_1d$. Then $r = a - bq = d(a_1 - b_1q) \Rightarrow d|r$.

3 Let $a = sd$, $b = td$. Then $ma + nb = d(ms + nt) \Rightarrow d|ma + nb$.

4 Let $c = bx$ and $b = ay$. Then $c = ayx \Rightarrow a|c$.

5 Let $b = ax$ and $a = by$. Then $a = axy \Rightarrow xy = 1 \Rightarrow x = 1$, $y = 1$ or $x = -1$, $y = -1 \Rightarrow a = \pm b$.

6 Let $a \in \mathbb{Z}$. Then by the division theorem, there exist unique integers q, r such that $a = 2q + r$, where $0 \le r < 1$.
Hence either $a = 2q$ or $a = 2q + 1$, for some $q \in \mathbb{Z}$.

7 By the division theorem, there exist unique integers q, r such that $n + b - 1 = bq + r$, where $0 \le r < b$.
Hence $n + (b - 1 - r) = bq$ and so the number $n + (b - 1 - r)$ is divisible by b for just one value of r, where $0 \le r < b$.
Putting r equal to each of the numbers $b - 1, b - 2, \ldots, 2, 1, 0$, in turn, shows that just one of the numbers $n, n + 1, \ldots, n + b - 1$ is divisible by b.

Problems 4.3a

1 (a) i.e. $1 = 5(3) + 7(-2) \Rightarrow k = 5(3k) + 7(-2k), \forall\ k \in \mathbb{Z}$.
(b) i.e. $1 = 8(5) + 13(-3) \Rightarrow k = 8(5k) + 13(-3k), \forall\ k \in \mathbb{Z}$.

2
$$
\begin{aligned}
38 &= 16(2) + 6 \Rightarrow 6 = 38 - 16(2) &&\ldots (1)\\
16 &= 6(2) + 4 \Rightarrow 4 = 16 - 6(2) &&\ldots (2)\\
6 &= 4(1) + 2 \Rightarrow 2 = 6 - 4(1). &&\ldots (3)\\
4 &= 2(2).
\end{aligned}
$$

Hence $\gcd(38, 16) = 2$ (the last non-zero remainder).
$$
\begin{aligned}
\text{Then } 2 &= 6 - 4(1), \quad \text{from (3)}\\
&= 6 - \{16 - 6(2)\}(1), \quad \text{from (2)}\\
&= 6(3) - 16(1)\\
&= \{38 - 16(2)\}(3) - 16(1), \quad \text{from (1)}\\
&= 38(3) - 16(7).
\end{aligned}
$$
Hence $2n = 38(3n) + 16(-7n), \forall\ n \in \mathbb{Z}$.

3
$$
\begin{aligned}
3059 &= 1547(1) + 1512 \Rightarrow 1512 = 3059 - 1547(1) &&\ldots (1)\\
1547 &= 1512(1) + 35 \Rightarrow 35 = 1547 - 1512(1) &&\ldots (2)\\
1512 &= 35(43) + 7 \Rightarrow 7 = 1512 - 35(43). &&\ldots (3)\\
35 &= 7(5).
\end{aligned}
$$

Hence $\gcd(3059, 1547) = 7$ (the last non-zero remainder).
$$
\begin{aligned}
7 &= 1512 - \{1547 - 1512(1)\}(43), \quad \text{from (3) and (2)}\\
&= 1512(44) - 1547(43)\\
&= \{3059 - 1547(1)\}(44) - 1547(43), \quad \text{from (1)}\\
&= 3059(44) - 1547(87).
\end{aligned}
$$
Hence $y = 44$ and $x = -87$.

4
$$
\begin{aligned}
1001 &= 30(33) + 11\\
30 &= 11(2) + 8\\
11 &= 8(1) + 3\\
8 &= 3(2) + 2\\
3 &= 2(1) + 1
\end{aligned}
$$

Hence $1 = 3-2(1) = 3 - \{8-3(2)\}(1) = 3(3) - 8 = \{11-8(1)\}(3) - 8$
$= 11(3) - 8(4) = 11(3) - \{30 - 11(2)\}(4) = 11(11) - 30(4)$
$= \{1001 - 30(33)\}(11) - 30(4)$
$= 1001(11) - 30(367),$

giving $s = 11$ and $t = -367$.

Let $k \in \mathbb{Z}$. Then $k = 1001(11k) + 30(-367k) \in \{1001s + 30t: s, t \in \mathbb{Z}\}$.

5 $19 = 15(1) + 4$
$15 = 4(3) + 3$
$4 = 3(1) + 1.$

Hence $1 = 4 - 3(1) = 4 - \{15 - 4(3)\}(1) = 4(4) - 15(1)$
$= \{19 - 15(1)\}(4) - 15(1) - 19(4) - 15(5).$

Thus $1 = 19x + 15y$, where $x = 4$, $y = -5$;

(a) $6 = 19(6x) + 15(6y) = 19(24) + 15(-30);$
(b) $6 = 6(19)x + 6(15)y = 114(4) + 90(-5);$
(c) $6 = 2(19)(3x) + 2(15)(3y) = 38(12) + 30(-15);$
$7 = 19(7x) + 15(7y) = 19(28) + 15(-35);$
$7 = 114x + 90y \Rightarrow 6|7.$ Hence 7 is not a linear combination of 114 and 90.
$7 = 38x + 30y \Rightarrow 2|7.$ Hence 7 is not a linear combination of 38 and 30.

6 $51 = 39(1) + 12,$
$39 = 12(3) + 3,$
$12 = 3(4).$

Hence $\gcd(51, 39) = 3$.

Now $3 = 39 - 12(3) = 39 - \{51 - 39(1)\}(3) = 39(4) - 51(3)$.

Hence $15 = 39(20) + 51(-15),$ giving $x = 20$, $y = -15$.

But $5 = 39x + 51y \Rightarrow 5 = 3(13x + 17y) \Rightarrow 3|5.$ Impossible!

7 $\gcd(a_1, b_1) = 1 \Rightarrow 1 = a_1x + b_1y,$ for some $x, y \in \mathbb{Z}$
$\Rightarrow d = da_1x + db_1y$
$= ax + by.$

Let $\gcd(a, b) = g$. Then $g|a$ and $g|b \Rightarrow g|d \Rightarrow g \le d$.
But $d|a$ and $d|b \Rightarrow d \le g$, by definition of g. Hence $d = g$.

Problems 4.4a

1 (a) $n \equiv 7 \pmod{12} \Rightarrow n = 7 + 12k = 3 + 4(1 + 3k) \Rightarrow n \equiv 3 \pmod 4;$
(b) $n \equiv 10 \pmod{18} \Rightarrow n = 10 + 18h = 1 + 3(3 + 6h)$
$\Rightarrow n \equiv 1 \pmod 3;$

(c) $n \equiv 13 \pmod{24} \Rightarrow n = 13 + 24t = 1 + 6(2 + 4t) \Rightarrow n \equiv 1 \pmod 6$.

2 Let $P(n)$ be the statement: $a \equiv b \pmod m \Rightarrow a^n \equiv b^n \pmod m$.
  (i) $P(1)$ true.
  (ii) Suppose $P(k)$ true for some $k \in \mathbb{Z}^+$. Then $a^k \equiv b^k \pmod m$.
      Hence, using result 4.4.1(d), with $c = a^k$, $d = b^k$, we have
      $aa^k \equiv bb^k \pmod m$.
      Thus $P(k)$ true $\Rightarrow P(k + 1)$ true and hence $P(n)$ is true,
      $\forall\, n \in \mathbb{Z}^+$.
  (a) $10 \equiv 1 \pmod 9 \Rightarrow 10^n \equiv 1^n \pmod 9 \Rightarrow 10^n \equiv 1 \pmod 9$.
  (b) $10 \equiv 1 \pmod 3 \Rightarrow 10^n \equiv 1^n \pmod 3 \Rightarrow 10^n \equiv 1 \pmod 3$.
  (c) $10 \equiv -1 \pmod{11} \Rightarrow 10^n \equiv (-1)^n \pmod{11} \Rightarrow 10^n \equiv 1 \pmod{11}$
      if n is even and $10^n \equiv -1 \pmod{11}$ if n is odd.
  (d) $5 \equiv 1 \pmod 4 \Rightarrow 5^n \equiv 1^n \pmod 4 \Rightarrow 5^n \equiv 1 \pmod 4$.
  (e) $9^2 \equiv 1 \pmod{10} \Rightarrow 9^{2n} \equiv 1^n \pmod{10} \Rightarrow 9^{2n} \equiv 1 \pmod{10}$.

3 (a) $8x \equiv 12 \pmod{20} \Leftrightarrow 8x = 12 + 20y$, for some $y \in \mathbb{Z}$,
      $\Leftrightarrow 2x = 3 + 5y \Leftrightarrow 2x \equiv 3 \pmod 5$.
      General solution (of both congruences): $x = 4 + 5k$, $k \in \mathbb{Z}$.
  (b) $9x \equiv 3 \pmod{15} \Leftrightarrow 9x = 3 + 15y$, for some $y \in \mathbb{Z}$,
      $\Leftrightarrow 3x = 1 + 5y \Leftrightarrow 3x \equiv 1 \pmod 5$.
      General solution: $x = 2 + 5k$, $k \in \mathbb{Z}$.
  (c) $10x \equiv 6 \pmod{24} \Leftrightarrow 10x = 6 + 24y$, for some $y \in \mathbb{Z}$,
      $\Leftrightarrow 5x = 3 + 12y \Leftrightarrow 5x \equiv 3 \pmod{12}$.
      General solution: $x = 3 + 12k$, $k \in \mathbb{Z}$.
  (d) $10x \equiv 15 \pmod{25} \Leftrightarrow 2x \equiv 3 \pmod 5$.
      General solution: $x = 4 + 5k$, $k \in \mathbb{Z}$.
  (e) $14x \equiv 35 \pmod{63} \Leftrightarrow 2x \equiv 5 \pmod 9$.
      General solution: $x = 7 + 9k$, $k \in \mathbb{Z}$.
  (f) $18x \equiv 6 \pmod{42} \Leftrightarrow 3x \equiv 1 \pmod 7$.
      General solution: $x = 5 + 7k$, $k \in \mathbb{Z}$.

4 (a) $\gcd(15, 21) = 3$. Since $3 | 6$, solution exists.
      $15x \equiv 6 \pmod{21} \Leftrightarrow 15x = 6 + 21y$, for some $y \in \mathbb{Z}$,
      $\Leftrightarrow 5x = 2 + 7y$. By inspection, least positive solution is $x = 6$ and
      hence general solution is $x = 6 + 7k$, $k \in \mathbb{Z}$.
  (b) $\gcd(15, 40) = 5$. Since $5 \nmid 6$, there is no solution.
  (c) $\gcd(15, 36) = 3$. Since $3 | 6$, solution exists.
      $15x \equiv 6 \pmod{36} \Leftrightarrow 15x = 6 + 36y$, for some $y \in \mathbb{Z}$,
      $\Leftrightarrow 5x = 2 + 12y$. Hence, least positive solution: $x = 10$.
      General solution: $x = 10 + 12k$.
  (d) $\gcd(15, 7) = 1$ and hence a solution exists.
      $15x \equiv 6 \pmod 7 \Leftrightarrow 15x = 6 + 7y$ for some $y \in \mathbb{Z}$.
      Since we are working modulo 7, one of the integers $0, 1, \ldots, 6$
      must be a solution. By trial, $x = 6$ is the least positive solution.
      General solution: $x = 6 + 7k$.

5 $63 = 11(5) + 8$ Hence
$11 = 8(1) + 3$ $1 = 3 - \{8-3(2)\}(1) = 3(3) - 8(1)$
$8 = 3(2) + 2$ $= \{11-8(1)\}(3) - 8(1) = 11(3) - 8(4)$
$3 = 2(1) + 1.$ $=11(3) - \{63 - 11(5)\}(4) = 11(23) - 63(4).$
Thus we have $1 = 11(23) - 63(4)$. We must rearrange this in the form
$11x = 1 + 63y$. This gives

$$11(23) = 1 + 63(4) \qquad \qquad \text{. . . (1)}$$

Thus $x = 23$ and $y = 4$ is a solution.

(a) General solution of $11x \equiv 1 \pmod{63}$: $x = 23 + 63k$.
(b) Multiplying through (1) by 5 gives

$$11(115) = 5 + 63(20).$$

Thus $x = 115$ is a solution of $11x \equiv 5 \pmod{63}$.
General solution: $x = 115 + 63k$.

6 $22 = 13(1) + 9$ Hence: $1 = 9 - 4(2)$
$13 = 9(1) + 4$ $= 9 - \{13 - 9\}(2) = 9(3) - 13(2)$
$9 = 4(2) + 1.$ $= \{22 - 13\}(3) - 13(2)$
                                             $= 22(3) - 13(5).$

$$\text{Now, } 1 = 22(3) - 13(5) \Leftrightarrow 13(-5) = 1 + 22(-3) \quad \text{. . . (1)}$$

Comparing this with $13x = 1 + 22y$, gives $x = -5$, $y = -3$.
(a) Since $x = -5$ is a solution, the general solution is $x = -5 + 22k$.
Thus the least positive solution is $x = 17$.
(b) Multiplying through (1) by 3 gives

$$13(-15) = 3 + 22(-9).$$

Thus $x = -15$ is a solution of $13x \equiv 3 \pmod{22}$.
General solution: $x = -15 + 22k$;
least positive solution: $x = -15 + 22 = 7$
(c) Multiplying through (1) by 10 gives

$$13(-50) = 10 + 22(-30).$$

Thus $x = -50$ is a solution of $13x \equiv 10 \pmod{22}$.
General solution: $x = -50 + 22k$;
least positive solution: $x = -50 + 22(3) = 16.$

7 $192 = 84(2) + 24$ \qquad\qquad\qquad\qquad\qquad\qquad\qquad\qquad\qquad\qquad \text{. . . (1)}
$84 = 24(3) + 12$
$24 = 12(2).$
Hence $\gcd(192, 84) = 12$. Since $12|48$, a solution exists.
From (1), $84(-2) = 24 + 192(-1)$
                       $84(-4) = 48 + 192(-2).$
Thus $x = -4$ is a solution of $84x \equiv 48 \pmod{192}$.

General solution: $x = -4 + \dfrac{192}{12}k = -4 + 16k$.

Hence least positive solution: $x = -4 + 16 = 12$.

Problems 4.5a

1 Let $P(n)$ be the statement of result 4.5.2. Then $P(1)$ is trivially true.
   Suppose $P(k)$ is true for some $k \in \mathbb{Z}^+$.
   Let $m = a_1 a_2 \ldots a_{k+1} = a_1 b$, where $b = a_2 a_3 \ldots a_{k+1}$, and p be a
   prime such that $p|m$. Then $p|m \Rightarrow p|a_1 b \Rightarrow p|a_1$ or $p|b$, by result 4.5.1.
   But b is the product of only k factors and hence, by the induction
   hypothesis, $p|b \Rightarrow p|a_i$ for some i, $2 \leq i \leq k + 1$. Hence $p|m \Rightarrow p|a_i$,
   for some i, $1 \leq i \leq k + 1$. Thus $P(k)$ true $\Rightarrow P(k + 1)$ true and hence
   $P(n)$ is true for all $n \in \mathbb{Z}^+$. Putting $a_1 = a_2 = \ldots = a_n = a$ gives
   $p|a^n \Rightarrow p|a$, $\forall\, n \in \mathbb{Z}^+$.

2 Suppose $\sqrt{10} = a/b$, where a/b is a fraction in its lowest terms. Then
   $10 = a^2/b^2 \Rightarrow 10b^2 = a^2$. Follow through the argument of theorem 4.5.5
   with $p = 2$ (or with $p = 5$) to show that a and b have a common factor
   of 2 (or of 5), contrary to hypothesis.

3 Let $m = ga_1 b_1 = ab_1 = a_1 b$. Hence $a|m$ and $b|m$ and so m satisfies
   condition (a). To show m satisfies condition (b), let $r \in \mathbb{Z}^+$ be any
   common multiple of a and b. Suppose $r = ax$.
   Then $b|r \Rightarrow b|ax \Rightarrow gb_1|ga_1 x \Rightarrow b_1|a_1 x \Rightarrow b_1|x$, since
   $\gcd(a_1, b_1) = 1$ (result 4.3.3).
   Let $x = b_1 y$. Then $r = ax = ab_1 y$ and hence $r \geq m$ and m satisfies
   (b).

5 BINARY OPERATIONS

5.1 BINARY OPERATIONS AND CLOSURE

Consider the set $\mathbb{R}$ of real numbers. We are familiar with the operations of arithmetic, such as addition, subtraction and multiplication, which combine a pair of numbers in $\mathbb{R}$ to make a third number, denoted of course by $x + y$, $x - y$ and xy or $x \times y$. That is, $x + y$, $x - y$ and $x \times y$ have a meaning, for every $x, y \in \mathbb{R}$. If we restrict our choice of x and y to the elements of the non−zero reals $\mathbb{R}^*$, then $x \div y$ also has a meaning for every $x, y \in \mathbb{R}^*$. The following definition is a generalization of these examples.

Binary operations

Let S be a non-empty set. Then $\circ$ is a *binary operation* on S if $x \circ y$ is defined for every pair of elements $x, y \in S$.

Closure

Let $\circ$ denote a binary operation on S. Then S is said to be *closed* with respect to $\circ$ if $x \circ y \in S$ for every pair of elements $x, y \in S$.

Examples 5.1.1

1. *Addition* is a *binary* operation on $\mathbb{Z}^+$, since $m + n$ is defined for every pair of numbers $m, n \in \mathbb{Z}^+$. Further, since $m + n \in \mathbb{Z}^+$, for every m, $n \in \mathbb{Z}^+$, the set $\mathbb{Z}^+$ is *closed* with respect to addition.

2. *Subtraction* is a *binary* operation on $\mathbb{Z}^+$, since $m - n$ is defined for all $m, n \in \mathbb{Z}^+$. But $m - n \notin \mathbb{Z}^+$ for every pair, $m, n \in \mathbb{Z}^+$ (for example, when $m = 1$ and $n = 3$) and hence $\mathbb{Z}^+$ is *not* closed with respect to subtraction.

3. *Division* is not a binary operation on the set $\mathbb{Z}$, since $a \div b$ is not defined when $b = 0$.

Notation

Let S be any set of numbers. We denote by S^* the subset of S consisting of all the *non-zero* numbers in S, that is

$$S^* = \{x \in S : x \neq 0\}.$$

When S is a set of real numbers, we denote by S^+ the subset of S consisting of all the strictly positive numbers in S. Thus

$$S^+ = \{x \in S : x > 0\}.$$

PROBLEMS 5.1a

1 Decide which of the following sets are *not* closed with respect to the binary operation given, stating a specific counter-example to justify your claim.

(a) $\mathbb{Z}$, $-$; (b) $\mathbb{Z}^+$, $\div$; (c) $\mathbb{Q}^*$, $\times$; (d) $\mathbb{Q}^*$, $\div$;
(e) $\mathbb{Z}^*$, $\div$; (f) $\mathbb{R}^+$, $+$; (g) $\mathbb{R}^+$, $\div$; (h) $\mathbb{R}^+$, $-$;
(i) $S = \{1, 2, 3, 6, 12\}$, $\times$; (j) $S = \{-2, -1, 0, 1, 2\}$, $+$.

2 A binary operation $\circ$ is defined on the set $\mathbb{Z}$ by

$$a \circ b = a + b - ab.$$

(a) Find

(i) $3 \circ 5$; (ii) $2 \circ 2$; (iii) $a \circ 0$.

(b) Decide which, if any, of the following subsets of $\mathbb{Z}$ are *not* closed with respect to $\circ$, giving a specific counter-example to support your assertion.

(i) $\mathbb{Z}$; (ii) $\mathbb{Z}^+$; (iii) $\mathbb{Z}^*$; (iv) $\{0, 1, 2\}$;
(v) $\{-1, 0, 1\}$.

3 An operation $\circ$ is defined on a set of numbers S by

$$x \circ y = \frac{xy}{x+y} .$$

Decide whether $\circ$ is a binary operation in each of the following cases, giving a specific counter-example where appropriate.

(a) $S = \mathbb{Z}^+$; (b) $S = \mathbb{Q}$; (c) $S = \mathbb{Q}^*$; (d) $S = \mathbb{Q}^+$.

4 Decide which, if any, of the following are *not* closed with respect to the operation given, giving a specific counter-example to support your assertion.

(a) $\mathbb{Z}$; $a \circ b = a^2 + b$. (b) $\mathbb{Z}$; $a \circ b = \dfrac{a(b+1)}{2}$.

(c) $\mathbb{Q}^+; x \circ y = \dfrac{x+1}{y+1}$.

(d) $\mathbb{Q}^+; x \circ y = x^2 + y^2 - 3xy$.

(e) $\mathbb{C}^*; w \circ y = wz$.

5 Let M be the set of all 2×2 matrices with *integer* entries. Show that M is closed with respect to (a) matrix addition; (b) matrix multiplication.

6 Let S be the set of all $m \times n$ matrices with real entries. What is the condition on m and n for matrix multiplication to be a binary operation on S?

7 Given two sets A and B, let S be the set of all mappings from A into B. Let $\circ$ be the operation 'composition of mappings'. Under which of the following conditions on A and B is $\circ$ a *binary* operation on S:
(a) $B = A$;   (b) $B \cap A = \varnothing$;   (c) $B \subseteq A$;   (d) $B \nsubseteq A$?

8 Which of the following sets of matrices are *closed* with respect to matrix multiplication? Justify your answers.

(a) $\left\{ \begin{bmatrix} 1 & x \\ 0 & 1 \end{bmatrix} : x \in \mathbb{R} \right\}$;

(b) $\left\{ \begin{bmatrix} x & 1 \\ 1 & 0 \end{bmatrix} : x \in \mathbb{R} \right\}$;

(c) $\left\{ \begin{bmatrix} x & y \\ 0 & 0 \end{bmatrix} : x, y \in \mathbb{R} \right\}$;

(d) $\left\{ \begin{bmatrix} x & 0 \\ 0 & 0 \end{bmatrix} : x \in \mathbb{R} \right\}$;

(e) $\left\{ \begin{bmatrix} 0 & x \\ 0 & 0 \end{bmatrix} : x \in \mathbb{R} \right\}$;

(f) $\left\{ \begin{bmatrix} x & y \\ 2x & 2y \end{bmatrix} : x, y \in \mathbb{R} \right\}$.

5.2 COMMUTATIVE AND ASSOCIATIVE PROPERTIES

Let S be a set and $\circ$ denote a binary operation on S.

Commutative property

The operation $\circ$ is said to be *commutative* if

$$a \circ b = b \circ a,$$

for *every* pair of elements a and b in S.

Examples 5.2.1

1. Addition and multiplication are commutative on the set $\mathbb{R}$. That is

$$x + y = y + x \quad \text{and} \quad xy = yx \quad \text{for all } x, y \in \mathbb{R}.$$

2. Subtraction is not commutative on the set $\mathbb{R}$, since $x - y \neq y - x$ for every choice of x and y in $\mathbb{R}$.
   Counter-example: If $x = 5$, $y = 1$, then $x - y = 4$ but $y - x = -4$.

3. Consider the binary operation $\circ$ defined on $\mathbb{Z}$ by

$$a \circ b = a + 2b.$$

Then $b \circ a = b + 2a$ and hence $a \circ b \neq b \circ a$ for all choices of a and b in $\mathbb{Z}$.
Counter-example: If $a = 1$ and $b = 5$, then $a \circ b = 11$ and $b \circ a = 7$.
Hence the operation $\circ$ is *not* commutative on $\mathbb{Z}$.

Associative property

The operation $\circ$ is said to be *associative* if

$$(a \circ b) \circ c = a \circ (b \circ c)$$

for *every* choice of elements a, b, $c \in S$.

Notice that we have not changed the *order* in which the elements a, b, c are to be combined; an operation is associative if we are allowed to *regroup* the elements, while keeping them in the same order.

Examples 5.2.2

1. Addition and multiplication are associated on $\mathbb{R}$.
   That is,

$$(x + y) + z = x + (y + z) \text{ and } (xy)z = x(yz)$$

for every x, y, $z \in \mathbb{R}$.

2. Subtraction is not associative on $\mathbb{R}$ since $(x - y) - z \neq x - (y-z)$ for every choice of x, y, $z \in \mathbb{R}$.
   Counter-example: If $x = 1$, $y = 2$, $z = 3$, then $(x-y)-z = -4$ but $x - (y-z) = 2$.

3. Consider the binary operation $\circ$ defined on $\mathbb{Z}$ by

$$a \circ b = a + 2b.$$

Then $(a \circ b) \circ c = (a + 2b) \circ c$
$\qquad\qquad = a + 2b + 2c.$
But $a \circ (b \circ c) = a \circ (b + 2c)$
$\qquad\qquad = a + 2(b + 2c)$
$\qquad\qquad = a + 2b + 4c.$

Thus $(a \circ b) \circ c \neq a \circ (b \circ c)$ for *every* choice of $a, b, c \in \mathbb{Z}$. Counter-example: Let $a = b = 0$ and $c = 1$. Then $(a \circ b) \circ c = 2$ and $a \circ (b \circ c) = 4$. Hence the operation $\circ$ is *not* associative on $\mathbb{Z}$.

General associativity

Let S be a set and $\circ$ be an associative binary operation on S. Then since for every choice of $x, y, z \in S$ it is true that $(x \circ y) \circ z = x \circ (y \circ z)$, we can dispense with the brackets and write either expression unambiguously as $x \circ y \circ z$.

Now consider the combination of four elements, say $w \circ x \circ y \circ z$. There are five ways of grouping these elements in order to find the meaning of $w \circ x \circ y \circ z$; these are:
$\{(w \circ x) \circ y\} \circ z$, $\{w \circ (x \circ y)\} \circ z$, $w \circ \{(x \circ y) \circ z\}$, $w \circ \{x \circ (y \circ z)\}$ and $(w \circ x) \circ (y \circ z)$. We will show that if $\circ$ is *associative* then these five expressions all give the same result.

First:

$$\begin{aligned}
\{(w \circ x) \circ y\} \circ z &= \{w \circ (x \circ y)\} \circ z \\
&= w \circ \{(x \circ y) \circ z), \quad \text{since } (w \circ b) \circ z = w \circ (b \circ z) \\
&= w \circ \{x \circ (y \circ z)\} \\
&= (w \circ x) \circ (y \circ z), \quad \text{since } w \circ (x \circ c) = (w \circ x) \circ c.
\end{aligned}$$

Hence we can also write any of these expressions unambiguously without brackets as $w \circ x \circ y \circ z$. In general, we can show that if $\circ$ is *associative*, then any combination of n elements, $n \in \mathbb{Z}$, can be written unambiguously without brackets as

$$x_1 \circ x_2 \circ x_3 \circ \quad \ldots \quad \circ x_n, \quad \forall \, x_i \in S.$$

This can be proved by induction.

PROBLEMS 5.2a

1 Decide which of the following binary operations are
  (i) commutative,
  (ii) associative on the sets given.
  Give a proof or a counter-example as appropriate.

(a) $+$ on $\mathbb{C}$;   (b) $\times$ on $\mathbb{C}$;
(c) $\div$ on $\mathbb{Q}$;   (d) $\circ$ on $\mathbb{Z}$, where $a \circ b = 2(a + b)$;
(e) $\circ$ on $\mathbb{Z}$, where $a \circ b = a^2 b^2$;   (f) $\circ$ on $\mathbb{Z}$, where $a \circ b = |a|$;
(g) $\circ$ on $\mathbb{Z}$, where $a \circ b = a + b + 2ab$.

2 Let S be the set of all $m \times n$ rectangular matrices with real entries. Prove that the operation of matrix addition is

(a) commutative on S; (b) associative on S.

3 Let M be the set of all 2×2 square matrices with real entries. Prove that the operation of matrix multiplication is

(a) associative on M;
(b) *not* commutative on M, giving a counter-example.

4 Let S be a non-empty set. Prove that both the binary operations $\cup$ and $\cap$ are

(a) commutative; (b) associative on the set $\mathscr{P}(S)$.

5 (a) Let $S = \{x, y\}$. Find examples of subsets $A, B, C \subseteq S$ (not necessarily distinct) such that

(i) $A - B \neq B - A$;
(ii) $A - (B - C) \neq (A - B) - C$.

(b) Repeat (a) for the set $S = \{x\}$.
What can be deduced about the operation *set difference* on the set $\mathscr{P}(S)$, where S is any non-empty set?

6 Let A be a non-empty set and let S denote the set of mappings of A into A. Let $\circ$ denote the operation *composition of mappings*.
Let $\alpha, \beta, \gamma \in S$ and $a \in A$. Show that

$$\{(\alpha \circ \beta) \circ \gamma\}(a) = \{\alpha \circ (\beta \circ \gamma)\}(a).$$

Deduce that $\circ$ is associative on S.
Show by giving a specific example of a set A and mappings $\alpha, \beta: A \to A$ such that $\alpha \circ \beta \neq \beta \circ \alpha$, that $\circ$ is not commutative on S.

7 Let S be a non-empty set and $\circ$ be a binary operation on S with the following properties:

(I) S is closed with respect to $\circ$; (II) $\circ$ is associative on S;
(III) $\circ$ is commutative on S.

Let X be any non-empty subset of S. For each of the following statements either (a) explain why it is true for all choices of X or (b) show by giving specific examples that it can be sometimes true and sometimes false, depending on the choice of X.

(a) $\circ$ is a binary operation on X; (b) X is closed with respect to $\circ$;
(c) $\circ$ is associative on X; (d) $\circ$ is commutative on X.

PROBLEMS 5.2b

1 Decide which of the following binary operations are
   (a) commutative;
   (b) associative on the sets given.
   Give a proof or counter-example, as appropriate.

   (a) $\div$ on $\mathbb{C}^*$;
   (b) $\circ$ on $\mathbb{Z}$ where $a \circ b = a^2b$;
   (c) $\circ$ on $\mathbb{Z}$, where $a \circ b = 2ab$;
   (d) $\circ$ on $\mathbb{Z}$, where $a \circ b = a + b - ab$;
   (e) $\circ$ on $\mathbb{Z}$, where $a \circ b = a^2 + b^2$;
   (f) $\circ$ on $\mathbb{Z}$ where $a \circ b = |ab|$.

2 Give an example of a set S and a binary operation $\circ$ on S such that:
   (a) $\circ$ is both commutative and associative;
   (b) $\circ$ is commutative but not associative;
   (c) $\circ$ is associative but not commutative;
   (d) $\circ$ is neither commutative nor associative.

3 The operation $\triangle$ is defined on the power set $\mathcal{P}(S)$ of a non-empty set S by $a \triangle B = A \cup B - A \cap B$. Prove that
   (a) $\triangle$ is commutative;   (b) $\triangle$ is associative.
   on the set $\mathcal{P}(S)$.

5.3 IDENTITY AND INVERSE ELEMENTS

Let S be a non-empty set and $\circ$ denote a binary operation on S.

Identity element

Suppose there exists an element $e \in S$ such that

$$e \circ x = x \circ e = x, \qquad \forall x \in S,$$

then e is called an *identity* element of S with respect to $\circ$. We say that the operation $\circ$ *admits* an identity element in S.

Examples 5.3.1

1. Since $1 \times a = a \times 1 = a$, $\forall a \in \mathbb{R}$, the number 1 is called the *identity* element of $\mathbb{R}$ with respect to *multiplication*, or *multiplicative identity*.

2. Since $0 + a = a + 0 = a$, $\forall\, a \in \mathbb{R}$, the number 0 is called the *identity* element of $\mathbb{R}$ with respect to *addition*, or *additive identity*.

3. Let $\circ$ be the operation defined on $\mathbb{Q}$ by

$$a \circ b = a + b + 2ab.$$

Suppose $\circ$ admits an identity element e. Then given any $a \in \mathbb{Q}$, e satisfies:

$$a \circ e = a, \qquad \forall\, a \in \mathbb{Q}$$
$$\Leftrightarrow a + e + 2ae = a, \qquad \forall\, a \in \mathbb{Q},$$
$$\Leftrightarrow e = 0, \qquad \text{since } 1 + 2a \neq 0 \text{ for every } a \in \mathbb{Q}.$$

Hence $a \circ 0 = a$, for all $a \in S$. But we must also check that

$$0 \circ a = a, \qquad \text{for all } a \in S.$$

Now, $\qquad\qquad 0 \circ a = 0 + a + 2 \times 0 \times a = a.$

Hence 0 is the identity element of $\mathbb{Q}$ with respect to $\circ$.

4. Let $\circ$ be the operation defined on $\mathbb{Z}$ by

$$a \circ b = ab^2.$$

Suppose $\circ$ admits an identity element $e \in \mathbb{Z}$. Then given any integer a, e satisfies:

$$a \circ e = a, \qquad \forall\, a \in \mathbb{Z}$$
$$\Leftrightarrow ae^2 = a, \qquad \forall\, a \in \mathbb{Z}$$
$$\Leftrightarrow e^2 = 1$$
$$\Leftrightarrow e = \pm 1.$$

Hence $a \circ 1 = a$ and $a \circ (-1) = a$, for all $a \in \mathbb{Z}$. But we must also check that $1 \circ a = a$ and $(-1) \circ a = a$, for all $a \in \mathbb{Z}$.
Now, $1 \circ a = 1 \times a^2 = a^2 \neq a$, for every $a \in \mathbb{Z}$
and $(-1) \circ a = (-1) \times a^2 = -a^2 \neq a$, for every $a \in \mathbb{Z}$.
Hence neither 1 nor -1 is an *identity* element of $\mathbb{Z}$ with respect to $\circ$.

Left and right identities

An element $a \in S$ is called a *left identity* with respect to $\circ$ if $e \circ a = a$, $\forall\, a \in S$, and is called a *right identity* if $a \circ e = a$, $\forall\, a \in S$.

Thus in the last example above, both 1 and -1 are *right identities*, but neither is a *left identity*.

These definitions imply that

$$e \text{ is an } identity \Leftrightarrow e \text{ is both a } left \text{ and a } right \text{ identity.}$$

131

RESULT 5.3.1   Let ∘ be a binary operation on a non-empty set S. Then ∘ admits *at most* one identity element on S.

Proof Suppose, if possible, that ∘ admits two identity elements, e and f.
Then $e \circ f = e$, since f is an identity.
and $e \circ f = f$, since e is an identity.
Hence $e = f$.
Thus if ∘ admits an identity, it is *unique*. ∎

Inverse elements

Let ∘ be a binary operation on S and suppose that ∘ admits an identity element e. Let $a\ e\ S$. If we can find an element $b\ e\ S$ such that

$$a \circ b = b \circ a = e$$

then b is called the *inverse* (element) of a, and we frequently write

$$b = a^{-1}.$$

Examples 5.3.2

1. Addition is a binary operation on $\mathbb{R}$ admitting the identity element 0. Let $x \in \mathbb{R}$. Then y is the inverse of x with respect to $+$ if

$$x + y = y + x = 0.$$

   Hence $y = (-x)$ is the inverse of x with respect to $+$.
   We call $-x$ the *additive* inverse of x.
   Since $-x \in \mathbb{R}$, $\forall\ x \in \mathbb{R}$, every element of $\mathbb{R}$ has an additive inverse in $\mathbb{R}$.

2. Multiplication is a binary operation on $\mathbb{R}$ admitting the identity element 1. Let $x \in \mathbb{R}$. Then y is the inverse of x with respect to multiplication if

$$x \times y = y \times x = 1.$$

   Hence if $x \neq 0$, $y = \dfrac{1}{x}$ is the inverse of x with respect to multiplication.

   We call $\dfrac{1}{x}$ the *multiplicative* inverse of x. The element $0 \in \mathbb{R}$ has no *multiplicative* inverse.

3 Consider the operation ∘ defined on Q by

$$a \circ b = a + b + 2ab.$$

We have shown in exercises 5.3.1 that ∘ admits the identity element 0. Let $a \in Q$. Then b is the inverse of a with respect to ∘ if

$$a \circ b = b \circ a = 0.$$

Now $\qquad a \circ b = 0 \Leftrightarrow a + b + 2ab = 0$

$$\Leftrightarrow b = \frac{-a}{1+2a}.$$

Since ∘ is commutative, $a \circ b = 0 \Rightarrow b \circ a = 0$.
Hence, provided $1 + 2a \neq 0$, i.e. $a \neq -\frac{1}{2}$. the element a has an inverse given by $a^{-1} = \dfrac{-a}{1+2a} \in Q$.

The element $-\frac{1}{2} \in Q$ has no inverse with respect to ∘.

PROBLEMS 5.3a

1 (a) Decide which of the following operations ∘ defined on the set Q of rationals admit an *identity* element in Q.

(i) $x \circ y = |x + y|$; (ii) $x \circ y = x + 2y$;
(iii) $x \circ y = x + y + 1$; (iv) $x \circ y = x$;
(v) $x \circ y = x + y - xy$; (vi) $x \circ y = x + (y+1)^2$.

(b) In the case of each of the operations in (a) that admit an identity in Q, find which elements of Q, if any, have *inverses* with respect to ∘.

2 (a) Let M be the set of all 2×2 matrices with entries in $\mathbb{R}$. Find the identity element of M with respect to matrix multiplication.

(b) Let H be the subset of M defined by $H = \left\{ \begin{bmatrix} x & 2x \\ 0 & 0 \end{bmatrix} : x \in \mathbb{R} \right\}$. Show that H also contains an identity element with respect to multiplication, but that the multiplicative identity of H is different from the multiplicative identity of M.

Find the multiplicative inverse of the matrix $\begin{bmatrix} x & 2x \\ 0 & 0 \end{bmatrix}$, where $x \neq 0$, in H.

Show that this matrix has no multiplicative inverse in M, for any value of x.

3 Let S be a non-empty set. Show that both the binary operations $\cup$ and $\cap$ admit an identity element in $\mathscr{P}(S)$. Given a subset $X \subseteq S$, investigate whether X has an inverse in $\mathscr{P}(S)$ with respect to (a) $\cup$ and (b) $\cap$.

4 Let A be a non-empty set and let S be the set of mappings of A into A. What is the identity element of S with respect to the operation *composition of mappings*?

PROBLEMS 5.3b

1 (a) Decide whether each of the following sets of matrices contains an identity element with respect to matrix multiplication.

  (i) $\left\{ \begin{bmatrix} 1 & x \\ 0 & 1 \end{bmatrix} : x \in \mathbb{R} \right\}$;   (ii) $\left\{ \begin{bmatrix} x & y \\ 0 & 0 \end{bmatrix} : x, y \in \mathbb{R} \right\}$;

  (iii) $\left\{ \begin{bmatrix} x & 0 \\ 0 & 0 \end{bmatrix} : x \in \mathbb{R} \right\}$;   (iv) $\left\{ \begin{bmatrix} x & -x \\ 2x & -2x \end{bmatrix} : x \in \mathbb{R} \right\}$.

  (b) In the case of each of the sets in (a) which possess a multiplicative identity, investigate which elements of the set have multiplicative inverses.

2 Let $J = \{a + ib: a, b \in \mathbb{Z}\}$.

  (a) Prove that J is closed with respect to multiplication of complex numbers.

  (b) Show that J contains a multiplicative identity.

  (c) Find *all* the elements of J which have multiplicative inverses *in J*.

3 Let $\circ$ denote a binary operation on a non-empty set S. Suppose that $\circ$ admits a left identity e and a right identity f. Prove that $e = f$.

4 Let $\circ$ denote an *associative* binary operation on a non-empty set S. Let $a \in S$ and suppose, if possible, that a has two inverses, say b_1 and b_2. By considering the combination $b_1 \circ a \circ b_2$, prove that $b_1 = b_2$

SOLUTIONS TO PROBLEMS IN THE (a) SETS

Problems 5.1a

  (b) 3, $5 \in \mathbb{Z}^+$, but $3 \div 5 \notin \mathbb{Z}^+$;   (e) 1, $2 \in \mathbb{Z}^*$, but $1 \div 2 \notin \mathbb{Z}^*$;
  (h) 2, $5 \in \mathbb{R}^+$, but $2 - 5 \notin \mathbb{R}^+$.   (i) 3, $6 \in S$, but $3 \times 6 \notin S$;
  (j) 1, $2 \in S$, but $1 + 2 \notin S$.

2 (a) (i) -7;   (ii) 0;   (iii) a.

    (b) (ii) 2, $3 \in \mathbb{Z}^+$, but $2 \circ 3 = -1 \notin \mathbb{Z}^+$;

       (iii) $2 \in \mathbb{Z}^*$, but $2 \circ 2$; $\notin \mathbb{Z}^*$.

       (v) $(-1) \circ (-1) = -3 \notin \{-1, 0, 1\}$.

3 (a) $x, y \in \mathbb{Z}^+ \Rightarrow x + y \neq 0$ for any $x, y \in \mathbb{Z}^+$. Hence $xy/(x + y)$ is defined for all $x, y \in \mathbb{Z}^+$. Thus $\circ$ is a *binary* operation on $\mathbb{Z}^+$.

    (b), (c) 1, $-1 \in \mathbb{Q}^*$, but $1 \circ (-1)$ is not defined. Hence $\circ$ is not *binary* on $\mathbb{Q}^*$ or $\mathbb{Q}$.

    (d) $x, y \in \mathbb{Q}^+ \Rightarrow x + y \neq 0$. Hence $xy/x+y$ is defined for all $x, y \in \mathbb{Q}^+$. Thus $\circ$ is a *binary* operation on $\mathbb{Q}^+$.

4 (b) 3, $2 \in \mathbb{Z}$ but $3 \circ 2 = 4\frac{1}{2} \notin \mathbb{Z}$;   (d) $1 \in \mathbb{Q}^+$, but
$$1 \circ 1 = (-1) \notin \mathbb{Q}^+.$$

5 Let $x_i = \begin{bmatrix} x_i & y_i \\ z_i & w_i \end{bmatrix}$, $i = 1, 2$. Then $X_1 + X_2 = \begin{bmatrix} x_1 + x_2 & y_1 + y_2 \\ z_1 + z_2 & w_1 + w_2 \end{bmatrix}$ and

$X_1 X_2 = \begin{bmatrix} x_1 x_2 + y_1 z_2 & x_1 y_2 + y_1 w_2 \\ z_1 x_2 + w_1 z_2 & z_1 y_2 + w_1 w_2 \end{bmatrix}$. Now if $x_i, y_i, z_i, w_i \in \mathbb{Z}$, then the entries of $X_1 + X_2$ and $X_1 X_2$ are all integers also. Hence M is closed with respect to (a) matrix addition; (b) matrix multiplication.

6 Let $A, B \in S$. Then AB is defined only when the number of columns of A = number of rows of B. Hence the condition for AB to be defined for all $A, B \in S$ is that $m = n$.

7 Let $\alpha, \beta \in S$. Then $\beta \circ \alpha$ is defined $\Leftrightarrow$ Im $\alpha \subseteq$ Dom β. Hence $\circ$ is binary if (a) $B = A$; (b) $B \subseteq A$.

8 Denote the given set of matrices by S, in each case.

    (a) Let $X_1 = \begin{bmatrix} 1 & x_i \\ 0 & 1 \end{bmatrix}$, $x_i \in \mathbb{R}$, $i = 1, 2$. Then $X_1, X_2 \in S$,

       and $X_1 X_2 = \begin{bmatrix} 1 & x_1 + x_2 \\ 0 & 1 \end{bmatrix} \in S$. Hence S is *closed*.

    (b) Let $X_i = \begin{bmatrix} x_i & 1 \\ 1 & 0 \end{bmatrix}$, $x_i \in \mathbb{R}$, $i = 1, 2$. Then $X_1, X_2 \in S$.

       But $X_1 X_2 = \begin{bmatrix} x_1 x_2 + 1 & x_1 \\ x_2 & 1 \end{bmatrix} \notin S$, for any $x_i \in \mathbb{R}$. Hence S is *not* closed with respect to matrix multiplication.

    (c) Let $X_i = \begin{bmatrix} x_i & y_i \\ 0 & 0 \end{bmatrix}$, $x_i, y_i \in \mathbb{R}$, $i = 1, 2$. Then $X_1, X_2 \in S$,

       and $X_1 X_2 = \begin{bmatrix} x_1 x_2 & x_1 y_2 \\ 0 & 0 \end{bmatrix} \in S$. Hence S is *closed*.

    (d) Let $X_1 = \begin{bmatrix} x_i & 0 \\ 0 & 0 \end{bmatrix}$, $x_i \in \mathbb{R}$, $i = 1, 2$. Then $X_1, X_2 \in S$, and

$$X_1X_2 = \begin{bmatrix} x_1x_2 & 0 \\ 0 & 0 \end{bmatrix} \in S. \text{ Hence } S \text{ is } closed.$$

(e) Let $X_i = \begin{bmatrix} 0 & x_i \\ 0 & 0 \end{bmatrix}$, $x_i \in \mathbb{R}$. Then X_1, $X_2 \in S$, and

$$X_1X_2 = \begin{bmatrix} 0 & 0 \\ 0 & 0 \end{bmatrix} \in S. \text{ Hence } S \text{ is } closed.$$

(f) Let $x_i = \begin{bmatrix} x_i & y_i \\ 2x_i & 2y_i \end{bmatrix}$, $x_i, y_i \in \mathbb{R}$, $i = 1, 2$. Then X_1, $X_2 \in S$ and

$$X_1X_2 = \begin{bmatrix} x_1x_2 + 2y_1x_2 & x_1y_2 + 2y_1y_2 \\ 2(x_1x_2 + 2y_1x_2) & 2(x_1y_2 + 2y_1y_2) \end{bmatrix} \in S.$$

Hence S is *closed*.

Problems 5.2a

1 (a) Let $z_k = x_k + iy_k$, $k = 1, 2$.

(i) Then $z_1 + z_2 = (x_1 + x_2) + i(y_1 + y_2)$
$= (x_2 + x_1) + i(y_2 + y_1) = z_2 + z_1$.
Hence + is *commutative*.

(ii) $z_1 + (z_2 + z_3) = x_1 + (x_2 + x_3) + i\{y_1 + (y_2 + y_3)\}$
$= (x_1 + x_2) + x_3 + i\{(y_1 + y_2) + y_3\}$
$= (z_1 + z_2) + z_3$.
Hence + is *associative* on $\mathbb{C}$.

(b) (i) $z_1z_2 = (x_1 + iy_1)(x_2 + iy_2) = x_1x_2 - y_1y_2 + i(x_1y_2 + y_1x_2)$.
$z_2z_1 = x_2x_1 - y_2y_1 + i(y_2x_1 + x_2y_1) = z_1z_2$.
Hence $\times$ is *commutative* on $\mathbb{C}$.

(ii) $z_1(z_2z_3) = (x_1 + iy_1)\{(x_2 + iy_2)(x_3 + iy_3)\}$
$= (x_1 + iy_1)\{x_2x_3 - y_2y_3 + i(y_2x_3 + x_2y_3)\}$
$= \{x_1x_2x_3 - x_1y_2y_3 - y_1y_2x_3 - y_1x_2y_3$
$+ i(y_1x_2x_3 - y_1y_2y_3 + x_1y_2x_3 + x_1x_2y_3)\}$.
$(z_1z_2)z_3 = \{x_1x_2 - y_1y_2 + i(x_1y_2 + y_1x_2)\}(x_3 + iy_3)$
$= \{x_1x_2x_3 - y_1y_2x_3 - x_1y_2y_3 - y_1x_2y_3$
$+ i(x_1y_2x_3 + y_1x_2x_3 + x_1x_2y_3 - y_1y_2y_3)\}$.
Hence $z_1(z_2z_3) = (z_1z_2)z_3$ and $\times$ is *associative* on $\mathbb{C}$.

(c) (i) $a \div b \neq b \div a$ for all choices of a, b. Counter-example: choose $a \neq b$.
Hence $\div$ is *not* commutative on $\mathbb{Q}$.

(ii) $a \div (b \div c) = ac/b$; $(a \div b) \div c = a/bc$.
Hence $a \div (b \div c) \neq (a \div b) \div c$ for all choices of a, b, c.
Counter-example: Choose $c \neq 1$.
Hence $\div$ is *not* associative on $\mathbb{Q}^*$.

(d) (i) $b \circ a = 2(b + a) = 2(a + b) = a \circ b$.
Hence $\circ$ is *commutative*.

(ii) $a \circ (b \circ c) = a \circ 2(b + c) = 2\{a + 2(b + c)\} = 2\{a + 2b + 2c\}$
$(a \circ b) \circ c = 2(a + b) \circ c = 2\{2(a + b) + c\} = 2\{2a + 2b + c\}$.
Hence $a \circ (b \circ c) \neq (a \circ b) \circ c$ for all choices of a, b and c.
Counter-example: Choose $a \neq c$.
Hence $\circ$ is *not* associative.

(e) (i) $b \circ c = b^2a^2 = a^2b^2 = a \circ b$;
Hence $\circ$ is *commutative*.

(ii) $a \circ (b \circ c) = a \circ (b^2c^2) = a^2b^4c^4$;
$(a \circ b) \circ c = (a^2b^2) \circ c = a^4b^4c^2$.
Hence $a \circ (b \circ c) \neq (a \circ b) \circ c$ for all choices of a, b, c.
Counter-example: Choose $a \neq c$.
Hence $\circ$ is *not* associative.

(f) (i) $a \circ b = |a|$ but $b \circ a = |b|$.
Hence $a \circ b \neq b \circ a$ for all choices of a and b.
Counter-example: Choose $a \neq b$.
Hence $\circ$ is *not* commutative.

(ii) $a \circ (b \circ c) = a \circ |b| = |a|$; $(a \circ b) \circ c = |a| \circ c = |a|$.
Hence $a \circ (b \circ c) = (a \circ b) \circ c$ and $\circ$ is *associative*.

(g) (i) $b \circ a = b + a + 2ba = a + b + 2ab = a \circ b$.
Hence $\circ$ is *commutative*.

(ii) $a \circ (b \circ c) = a \circ (b + c + 2bc)$
$= a + (b + c + 2bc) + 2a(b + c + 2bc)$
$= a + b + c + 2ab + 2bc + 2ca + 4abc$.
$(a \circ b) \circ c = (a + b + 2ab) \circ c$
$= (a + b + 2ab) + c + 2(a + b + 2ab)c$
$= a + b + c + 2ab + 2bc + 2ca + 4abc$.

Hence $(a \circ b) \circ c = a \circ (b \circ c)$ and $\circ$ is *associative*.

2 Let $A = (a_{ij})$, $B = (b_{ij})$, $C = (c_{ij})$.
(a) $A + B = (a_{ij} + b_{ij}) = (b_{ij} + a_{ij}) = B + A$.
Hence matrix addition is commutative.
(b) $(A + B) + C = (\{a_{ij} + b_{ij}\} + c_{ij}) = (a_{ij} + b_{ij} + c_{ij})$
$A + (B + C) = (a_{ij} + \{b_{ij} + c_{ij}\}) = (a_{ij} + b_{ij} + c_{ij})$.
Hence matrix addition is associative.

3 (a) Let $X_i = \begin{bmatrix} x_i & y_i \\ z_i & w_i \end{bmatrix}$, $i = 1, 2, 3$. Verify that $(X_1X_2)X_3$ and $X_1(X_2X_3)$
both give:

$$\begin{bmatrix} x_1x_2x_3 + y_1z_2x_3 + x_1y_2z_3 + y_1w_2z_3 & x_1x_2y_3 + y_1z_2y_3 + x_1y_2w_3 + y_1w_2w_3 \\ z_1x_2x_3 + w_1z_2x_3 + z_1y_2z_3 + w_1w_2z_3 & z_1x_2y_3 + w_1z_2y_3 + z_1y_2w_3 + w_1w_2w_3 \end{bmatrix}$$

(b) *Let* $X = \begin{bmatrix} 1 & 1 \\ 1 & 1 \end{bmatrix}$, $Y = \begin{bmatrix} 1 & 2 \\ 3 & 4 \end{bmatrix}$ (for example) and show $XY \neq YX$.

4 Let A, B, C be subsets of S.
(a) $A \cup B = \{x \in S: x \in A \text{ or } x \in B\} = \{x \in S: x \in B \text{ or } x \in A\}$
$$= B \cup A.$$
$A \cap B = \{x \in S: x \in A \text{ and } x \in B\} = \{x \in S: x \in B \text{ and } x \in A\}$
$$= B \cap A.$$
Hence both $\cup$ and $\cap$ are commutative.
(b) $A \cup (B \cup C) = \{x \in S: x \in A \text{ or } x \in B \cup C\}$
$$= \{x \in S: x \in A \text{ or } x \in (B \text{ or } C)\}$$
$$= \{x \in S: x \in (A \text{ or } B) \text{ or } x \in C\}$$
$$= \{x \in S: x \in A \cup B \text{ or } x \in C\}$$
$$= (A \cup B) \cup C.$$
A similar proof with 'and' instead of 'or' shows that $(A \cap B) \cap C = A \cap (B \cap C)$.
Hence $\cup$ and $\cap$ are both associative.

5 (a) (i) Let $A = \{x\}$, $B = \{y\}$. Then $A - B = \{x\}$ and $B - A = \{y\}$.
(ii) Let $A = \{x, y\}$, $B = \{x\}$, $C = \{y\}$.
Then $A - (B - C) = A - \{x\} = \{y\}$.
But $(A - B) - C = \{y\} - C = \emptyset$.
(b) (i) Let $A = \{x\}$ and $B = \emptyset$. Then $A - B = \{x\}$, $B - A = \emptyset$.
(ii) Let $A = C = \{x\}$ and $B = \emptyset$.
Then $A - (B - C) = A - \emptyset = \{x\}$.
But $(A - B) - C = \{x\} - C = \emptyset$.
Set difference is neither commutative nor associative.

6 Let $\gamma(a) = z$, $\beta(z) = y$.
Then $\{(\alpha \circ \beta) \circ \gamma\}(a) = (\alpha \circ \beta)(\gamma(a)) = (\alpha \circ \beta)(z) = \alpha(\beta(z)) = \alpha(y)$.
Now $\{\alpha \circ (\beta \circ \gamma)\}(a) = \alpha\{(\beta \circ \gamma)(a)\}$;
but $(\beta \circ \gamma)(a) = \beta(\gamma(a)) = \beta(z) = y$.
Hence $\{\alpha \circ (\beta \circ \gamma)\}(a) = \alpha(y) = \{(\alpha \circ \beta) \circ \gamma\}(a)$, and $\circ$ is *associative*.
Let $A = \{1, 2, 3\}$. Define α, $\beta: A \to A$ by: $\alpha(1) = 1$, $\alpha(2) = 3$, $\alpha(3) = 2$; $\beta(1) = 2$, $\beta(2) = 1$, $\beta(3) = 3$.
Then $(\beta \circ \alpha)(1) = \beta(1) = 2$, but $(\alpha \circ \beta)(1) = \alpha(2) = 3$.
Hence $\beta \circ \alpha \neq \alpha \circ \beta$ and $\circ$ is not *commutative*.
Or let $A = \mathbb{R}$. Define α, $\beta: \mathbb{R} \to \mathbb{R}$ by $\alpha(x) = x + 2$, $\beta(x) = x^2$.
Then $(\alpha \circ \beta)(x) = \alpha(x^2) = x^2 + 2$,
but $(\beta \circ \alpha)(x) = \beta(x + 2) = x^2 + 4x + 4$.

7 (a) True; $x, y \in X \Rightarrow x, y \in S \Rightarrow x \circ y$ is defined.
(b) Not necessarily true. Let $S = \mathbb{Z}$, $X = \mathbb{Z}^+$ and $\circ$ denote subtraction. Then $\mathbb{Z}$ is closed with respect to subtraction, but $\mathbb{Z}^+$ is not.
(c) True; $x, y, z \in X \Rightarrow x, y, z \in S \Rightarrow x \circ (y \circ z) = (x \circ y) \circ z$, by (II).
(d) True; $x, y \in X \Rightarrow x, y \in S \Rightarrow x \circ y = y \circ x$, by (III).

Problems 5.3a

1 (a) (i) no identity; (ii) no identity; (iii) -1;
  (iv) no identity; (v) 0; (vi) no identity.

  (b) (iii) Let y be the inverse of x.
    Then $x \circ y = -1 \Rightarrow x + y + 1 = -1 \Rightarrow y = -2 - x$.
    Hence every $x \in \mathbb{Q}$ has an inverse given by $-2 - x \in \mathbb{Q}$.

  (v) Let y be the inverse of x.
    Then
$$x \circ y = 0 \Rightarrow x + y - xy = 0 \Rightarrow y = \frac{x}{x-1} \Rightarrow \text{all elements}$$

  of $\mathbb{Q}$ have an inverse in $\mathbb{Q}$, except $x = 1$, which has no inverse.

2 (a) Multiplicative identity of M: $I = \begin{bmatrix} 1 & 0 \\ 0 & 1 \end{bmatrix}$.

  (b) Suppose $E = \begin{bmatrix} e & 2e \\ 0 & 0 \end{bmatrix}$ is multiplicative identity of H. Let

$$X = \begin{bmatrix} x & 2x \\ 0 & 0 \end{bmatrix} \in H.$$

  Then $EX = X \Rightarrow \begin{bmatrix} ex & 2ex \\ 0 & 0 \end{bmatrix} = \begin{bmatrix} x & 2x \\ 0 & 0 \end{bmatrix}, \forall x \in \mathbb{R} \Rightarrow e = 1.$

  Hence $E = \begin{bmatrix} 1 & 2 \\ 0 & 0 \end{bmatrix}$ is a left-identity of H. But $XE = \begin{bmatrix} x & 2x \\ 0 & 0 \end{bmatrix} = X$.
  Hence E is also a right-identity and so is the multiplicative identity of H.

  Let $Y = \begin{bmatrix} y & 2y \\ 0 & 0 \end{bmatrix}$ be the multiplicative inverse of X.

  Then $XY = E \Rightarrow \begin{bmatrix} xy & 2xy \\ 0 & 0 \end{bmatrix} = \begin{bmatrix} 1 & 2 \\ 0 & 0 \end{bmatrix} \Rightarrow y = \frac{1}{x}$, when $x \neq 0$.

  It is easy to verify that $YX = E$ when $y = \frac{1}{x}$, $x \neq 0$. Hence every non-zero matrix in H has a multiplicative inverse in H. Now the multiplicative identity of M is not E, but I. It is easy to check that $XY = I$ is impossible for any $Y \in M$.

3 Let $X \subseteq S$. Then $\varnothing \cup X = X \cup \varnothing = X$. Hence $\varnothing$ is the identity of $\mathcal{P}(S)$ with respect to $\cup$. Also $S \cap X = X \cap S = X$ and hence S is the identity of $\mathcal{P}(S)$ with respect to $\cap$.

  (a) If $Y \subseteq S$ is the inverse of X with respect to $\cup$, then $X \cup Y = \varnothing$, which is impossible unless $X = Y = \varnothing$. Hence no element of $\mathcal{P}(S)$ except $\varnothing$ has an inverse with respect to $\cup$.

  (b) If $Y \subseteq S$ is the inverse of X with respect to $\cap$, then $X \cap Y = S$ which is impossible unless $X = Y = S$. Hence no element of $\mathcal{P}(S)$ except S has an inverse with respect to $\cap$.

4 The identity mapping ι_A defined by $\iota_A(a) = a$, $\forall a \in A$.

6 GROUPS: SOME GROUNDWORK

6.1 AXIOMS AND EXAMPLES

Algebra is the study of abstract systems. One of the most fundamental of these systems is called a *group*. The notion of a group is used in many branches of mathematics and in a wide variety of applications, including computer design and programming languages, coding, elementary particle and nuclear physics, quantum mechanics, molecular structure and crystallography.

Axioms

Let G be a non-empty set on which a binary operation $\circ$ is defined. Suppose that

Axiom 1 G is *closed* with respect to $\circ$; i.e. $a \circ b \in G$, $\forall\ a, b \in G$.

Axiom 2 $\circ$ is *associative*; i.e. $(a \circ b) \circ c = a \circ (b \circ c)$, $\forall\ a, b, c \in G$.

Axiom 3 $\circ$ admits an *identity element* in G; i.e. we can find an element $e \in G$ such that $a \circ e = e \circ a = a$, $\forall\ a \in G$.

Axiom 4 Every element of G has an *inverse element* in G with respect to $\circ$; i.e. for each $a \in G$, there is an element $b \in G$ such that $a \circ b = b \circ a = e$.
Then G is called a *group* with respect to the binary operation $\circ$ and denoted by $(G, \circ)$.

Note: We do *not* require that $\circ$ is *commutative*. If, in addition to satisfying axioms 1 to 4, $\circ$ is also commutative, then we call $(G, \circ)$ an *abelian* group.

Examples 6.1.1

1. (a) Consider first the set $\mathbb{Z}^+$ with respect to addition. Axioms 1 and 2 are clearly satisfied; however, $(\mathbb{Z}^+, +)$ is *not* a group because $\mathbb{Z}^+$ does not contain an identity element with respect to addition and hence it is also impossible to define the additive inverse of each element of $\mathbb{Z}^+$. Thus axioms 3 and 4 are not satisfied.

   (b) Instead, let us consider $(\mathbb{Z}, +)$. Axioms 1 and 2 are satisfied. Further, $0 \in \mathbb{Z}$ and hence axiom 3 is also satisfied. Lastly, $\forall\, x \in \mathbb{Z}$, it is true that $-x \in \mathbb{Z}$ and $x + (-x) = (-x) + x = 0$. Hence each element of $\mathbb{Z}$ has an additive inverse in $\mathbb{Z}$ and axiom 4 is also satisfied. Thus $(\mathbb{Z}, +)$ is a group.

2. (a) Consider $(\mathbb{Z}, \times)$. Clearly axioms 1 and 2 are satisfied, and, since $1 \in \mathbb{Z}$ is a multiplicative identity, axiom 3 is also satisfied. However, given $a \in \mathbb{Z}$, it is not always true that its multiplicative inverse $\dfrac{1}{a} \in \mathbb{Z}$; consider for example $a = 2, -3, 4$ etc. and also $a = 0$ which has no multiplicative inverse, even in $\mathbb{R}$. Hence $(\mathbb{Z}, \times)$ is *not* a group.

   (b) If we consider $(\mathbb{Q}, \times)$ instead then we have almost overcome this problem since if $x \in \mathbb{Q}$, $x \neq 0$, then $\dfrac{1}{x} \in \mathbb{Q}$. However, $0 \in \mathbb{Q}$ and 0 has no multiplicative inverse in $\mathbb{Q}$. Thus not *every* element in $\mathbb{Q}$ has a multiplicative inverse in $\mathbb{Q}$. Hence $(\mathbb{Q}, \times)$ is *not* a group.

   (c) The way round this final difficulty is to exclude 0 from $\mathbb{Q}$ and consider instead $(\mathbb{Q}^*, \times)$. It is easy to see that all four axioms are satisfied by $(\mathbb{Q}^*, \times)$ and hence it is a group.

3. (a) Consider the set S of all square 2×2 matrices with real entries, with respect to matrix multiplication. If $A, B \in S$, then their product AB is also a 2×2 matrix with real entries and hence axiom 1 is satisfied. We shall simply state that matrix multiplication is associative. (For a proof see, for example, D. Towers, *Linear Algebra* (Macmillan, 1988).) The unit matrix $I = \begin{pmatrix} 1 & 0 \\ 0 & 1 \end{pmatrix} \in S$ and is easily shown to be a multiplicative identity element for S. However, not every matrix in S has a multiplicative inverse. Let $A = \begin{pmatrix} a & b \\ c & d \end{pmatrix} \in S$. Then we can show that there is a matrix $X \in S$ such that $AX = XA = I$ if and only if $ad-bc \neq 0$. (The matrices for which $ad-bc = 0$ are called *singular*. The others are called *nonsingular* or *invertible*.) Thus S is *not* a group with respect to matrix multiplication because axiom 4 is not satisified.

(b) Suppose instead we consider the subset S' of S which consists of all the *invertible* matrices of S and excludes the singular ones. Before leaping to the conclusion that S' is a group because we have thrown out the elements without inverses, it is necessary to check that S' is *closed*. For, suppose $A, B \in S'$. We know that $AB \in S$ but does $AB \in S'$, for every choice of A and B in S'? Now $A, B \in S' \Rightarrow A, B$ are invertible $\Rightarrow A^{-1}, B^{-1} \in S \Rightarrow B^{-1} A^{-1} \in S$, i.e. the matrix product $B^{-1} A^{-1}$ is a 2×2 matrix with real entries. But matrix multiplication is associative, so that $(AB)(B^{-1} A^{-1}) = A(BB^{-1}) A^{-1} = AIA^{-1} = AA^{-1} = $ I. Similarly, $(B^{-1} A^{-1})(AB) = I$ and hence AB is an invertible matrix. Thus $A, B \in S' \Rightarrow AB \in S'$ and S' is *closed* with respect to matrix multiplication. Thus the set S' of all 2×2 *invertible* matrices with real entries forms a group with respect to matrix multiplication. This group is known as *the general linear group of degree 2 over* $\mathbb{R}$ and is denoted by $GL(2, \mathbb{R})$.

4. Consider the set $\mathbb{R}^n = \{(a_1, a_2, \ldots, a_n): a_i \in \mathbb{R}\}$ of all real n-tuples. We can define an operation of addition on $\mathbb{R}^n$ by $(a_1, a_2, \ldots, a_n) + (b_1, b_2, \ldots, b_n) = (a_1 + b_1, a_2 + b_2, \ldots, a_n + b_n)$. Then it is easy to check that axioms 1 and 2 are satisfied. Also, $\mathbf{0} = (0, 0, \ldots, 0) \in \mathbb{R}^n$ and clearly is an additive identity element. Thus axiom 3 is also satisfied. Finally, given any element $\mathbf{a} = (a_1, a_2, \ldots, a_n) \in \mathbb{R}^n$, the element $\mathbf{a}' = (-a_1, -a_2, \ldots, -a_n) \in \mathbb{R}^n$ and satisfies $\mathbf{a} + \mathbf{a}' = \mathbf{a}' + \mathbf{a} = \mathbf{0}$. Hence axiom 4 is also satisfied and $(\mathbb{R}^n, +)$ is a group.

Note: In the following sets of problems, and in subsequent sets, you may *quote*, without proof, that the following operations are *associative*:

(a) addition and multiplication on any set of real or complex numbers;
(b) addition and multiplication of matrices;
(c) composition of mapping;
(d) set union and set intersection.

PROBLEMS 6.1.1a

1 Decide which of the following pairs of set and operation form a group. In the case of those that do not, give the axioms which are not satisfied.

(a) $(\mathbb{Q}, +)$; (b) $(\mathbb{Q}^+, \times)$; (c) $(\mathbb{C}, +)$; (d) $(\mathbb{R}^+, +)$; (e) $(\mathbb{Z}^+, \times)$;
(f) the set $3\mathbb{Z} = \{3a: a \in \mathbb{Z}\}$, with respect to addition.
(g) the set of all 2×2 matrices with integer entries, with respect to matrix addition;

(h) the set $S = \{2^r: r \in \mathbb{Z}\}$, with respect to multiplication;

(i) the power set $\mathcal{P}(A)$ of a non-empty set A, with respect to set union;

(j) the power set $\mathcal{P}(A)$ of a non-empty set A, with respect to set intersection.

2 Let A be the set $\mathbb{Q} - \{1\}$. Prove that A is a group with respect to the operation $\circ$ defined by

$$a \circ b = a + b - ab.$$

3 Prove that $\mathbb{Z}$ is a group with respect to the operation $\circ$ defined by

$$a \circ b = a + b + 1.$$

4 Prove that the following sets of matrices are groups with respect to matrix multiplication:

(a)
$$W_1 = \left\{ \begin{pmatrix} x & 0 \\ 0 & 0 \end{pmatrix} : x \in \mathbb{Q}^* \right\};$$

(b)
$$W_2 = \left\{ \begin{pmatrix} x & -x \\ 0 & 0 \end{pmatrix} : x \in \mathbb{Q}^* \right\}.$$

5 Let $S = \{(x, y): x \in \mathbb{R}, y \in \mathbb{R}^*\}$. An operation $\circ$ is defined on S by

$$(x_1, y_1) \circ (x_2, y_2) = (x_1 + x_2, y_1 y_2).$$

Prove that $(S, \circ)$ is a group. Give its identity element and the inverse of a typical element.

Finite and infinite groups

All the examples of groups considered so far have an infinite number of elements and are therefore called *infinite* groups. We next consider some examples of *finite* groups.

Cayley table

To check that a small finite set S is a group with respect to a given operation $\circ$ it is often convenient to construct a table, known as a *Cayley* table, after the British mathematician Arthur Cayley (1821–95). The elements of S are listed vertically down the left-hand side of the table and horizontally across the top of the table. The element $a \circ b$ is entered in the *row opposite a* and the *column below b*. It is easy to check from the table whether axioms 1, 3 and 4 are satisfied. It is not usually practical to verify the associative law from the table (see the paragraph on the associative law after example 6.1.2 below).

Example 6.1.2

Show that the set $H = \{1 + 0i, -1 + 0i, 0 + 1i, 0 - 1i\}$ forms a group with respect to multiplication of complex numbers.

Solution We first construct a Cayley table as described above. In so doing it will be convenient to refer to the elements of H as $1, -1, i, -i$, as this makes the table simpler to read.

$\times$	1	-1	i	$-i$
1	1	-1	i	$-i$
-1	-1	1	$-i$	i
i	i	$-i$	-1	1
$-i$	$-i$	i	1	-1

Notice that the group operation is usually indicated in the top left-hand corner of the table.

Axiom 1 Since all the *entries* in the table are elements of H, we have shown that $a \circ b \in H$, $\forall\, a, b \in H$.

Axiom 2 Multiplication of complex numbers is associative.

Axiom 3 From an inspection of the row opposite the element 1 and the column below the element 1 we see that $1 \circ a = a$, $\forall\, a \in H$; and $a \circ 1 = a$, $\forall\, a \in H$. Thus 1 is the identity element and axiom 3 is satisfied.

Axiom 4 From the table we can see that

$$1 \circ 1 = 1, \quad (-1) \circ (-1) = 1, \quad i \circ (-i) = 1 \quad \text{and} \quad (-i) \circ i = 1.$$

Thus each element of H has an inverse in H. Hence H is a group with respect to multiplication of complex numbers.

Associative law

When the operation $\circ$ is defined on a set S by an abstract formula for $a \circ b$ (such as 'addition mod m' or by a formula such as $a \circ b = a + b - ab$, etc.) it is always best to check the associative law directly from the formula and *not* from the Cayley table. This is because there are n^3 combinations of the form $a \circ b \circ c$, where $|S| = n$, and hence it would involve $2n^3$ calculations to verify that $(a \circ b) \circ c = a \circ (b \circ c)$ in each case!

Order of a finite group

Let G be a finite group. We define the *order* of G to be the number of elements in G. We use the same notation as for the number of elements in a set and denote the *order of G* by $|G|$.

So far, we only know of the existence of one finite group. We next look at the construction of a whole class of examples of finite groups. In fact we show that given any $n \in \mathbb{Z}^+$, it is always possible to find an example of a group of order n.

The sets $\mathbb{Z}_n$

In section 2.4, we defined the *congruence class* (or *residue class*) $\bar{a}$ of an integer a modulo n, for some given $n \in \mathbb{Z}^+$. That is,

$$\bar{a} = \{x \in \mathbb{Z}: x \equiv a \pmod{n}\}.$$

We showed that given any integer $x \in \mathbb{Z}$, then x belongs to exactly one of the congruence classes $\bar{0}, \bar{1}, \ldots, \overline{n-1} \pmod{n}$. That is the same as saying that x is congruent to just one of the integers $0, 1, 2, \ldots, n-1 \pmod{n}$. We define $\mathbb{Z}_n = \{\bar{0}, \bar{1}, \ldots, \overline{n-1}\}$.

Examples 6.1.3

Suppose $n = 5$. Then $\mathbb{Z}_5 = \{\bar{0}, \bar{1}, \bar{2}, \bar{3}, \bar{4}\}$.
Check each of the following statements.

(a) $9 \in \bar{4}$;   (b) $17 \in \bar{2}$;   (c) $-2 \in \bar{3}$;   (d) $125 \in \bar{0}$;
(e) $\bar{9} = \bar{4}$;   (f) $\overline{36} = \bar{1}$;   (g) $\overline{(-11)} = \bar{4}$;   (h) $\overline{35} = \bar{0}$.

Addition in $\mathbb{Z}_n$

We can define an additive operation $\oplus$ in $\mathbb{Z}_n$ by the rule

$$\bar{a} \oplus \bar{b} = \overline{a + b}.$$

Example 6.1.4

Consider $\mathbb{Z}_5$ again. The rule gives

(a) $\bar{2} \oplus \bar{1} = \bar{3}$;   (b) $\bar{3} \oplus \bar{4} = \bar{7} = \bar{2}$;   (c) $\bar{2} \oplus \bar{3} = \bar{5} = \bar{0}$;

(d) $\overline{17} \oplus \overline{36} = \overline{53} = \overline{3}$;

(e) $\overline{(-2)} \oplus \overline{19} = \overline{17} = \overline{2}$;

(f) $\overline{(-13)} \oplus \overline{(-2)} = \overline{(-15)} = \overline{0}$.

A problem arises with an operation defined in this way. The elements $\bar{a}$ and $\bar{b}$ of $\mathbb{Z}_n$ are themselves *sets* and we have therefore defined the addition of two sets in terms of the sum of a *representative* of each set. The question arises: suppose we had selected a different class representative of $\bar{a}$ and $\bar{b}$, would we have necessarily arrived at the same answer for $\bar{a} \oplus \bar{b}$? If this is so, the operation $\oplus$ is said to be *well-defined*.

In example 6.1.4, compare (a) with (d).

Here, $\overline{17} = \overline{2}$ and $\overline{36} = \overline{1}$ and we see that $\overline{17} \oplus \overline{36} = \overline{2} \oplus \overline{1}$. Comparing (b) with (e) and (c) with (f), we again see that by adding different representatives of the congruence classes we have achieved the same sum. We now prove that this will be true in general.

RESULT 6.1.1   The operation $\oplus$ on $\mathbb{Z}_n$ is well-defined.

Proof   Let $\bar{a}, \bar{b} \in \mathbb{Z}_n$. By the definition of $\oplus$, we have

$$\bar{a} \oplus \bar{b} = \overline{a + b}.$$

Now suppose $a_1 \in \bar{a}$ and $b_1 \in \bar{b}$. Then we could have chosen a_1 and b_1 as class representatives of $\bar{a}$ and $\bar{b}$ respectively. In this case, we would have

$$\bar{a} \oplus \bar{b} = \bar{a}_1 + \bar{b}_1 = \overline{a_1 + b_1}.$$

Thus it is necessary to show that $\overline{a_1 + b_1} = \overline{a + b}$.

Now $a_1 \in \bar{a} \Rightarrow a_1 \equiv a \pmod{n} \equiv a_1 = a + kn$, for some $k \in \mathbb{Z}$;

and $b_1 \in \bar{b} \Rightarrow b_1 \equiv b \pmod{n} \Rightarrow b_1 = b + hn$, for some $h \in \mathbb{Z}$.

Thus $a_1 + b_1 = a + b + (k + h)n$

$$\Rightarrow a_1 + b_1 \equiv a + b \pmod{n}$$

$$\Rightarrow \overline{a_1 + b_1} = \overline{a + b}.$$

This shows that $\oplus$ is well-defined. ∎

RESULT 6.1.2 $(\mathbb{Z}_n, \oplus)$ is an abelian group.

Proof

Axiom 1 Let $\bar{a}, \bar{b} \in \mathbb{Z}_n$. Then $\bar{a} \oplus \bar{b} = \overline{a + b}$, by definition. By the division theorem, we can find integers q, r, where $0 \leqslant r < n$ such that $a + b = qn + r$. Then $a + b \equiv r(\mathrm{mod}\ n)$ and hence $\overline{a + b} = \bar{r}$. But $\bar{r} \in \mathbb{Z}_n$ since $0 \leqslant r \leqslant n - 1$. Thus $\bar{a} \oplus \bar{b} \in \mathbb{Z}_n$, for all $\bar{a}, \bar{b} \in \mathbb{Z}_n$.

Axiom 2 We must show that
$$\bar{a} \oplus (\bar{b} \oplus \bar{c}) = (\bar{a} \oplus \bar{b}) \oplus \bar{c}, \ \forall \ a, b, c \in \mathbb{Z}.$$

Now $(\bar{a} \oplus (\bar{b} \oplus \bar{c}) = \bar{a} \oplus \overline{b + c} = \overline{a + (b + c)};$

and $(\bar{a} \oplus \bar{b}) \oplus \bar{c} = \overline{a + b} \oplus c = \overline{(a + b) + c}.$

But $(a + b) + c = a + (b + c),$

since addition in $\mathbb{Z}$ is associative. Thus $\oplus$ is associative.

Axiom 3 Since
$$\bar{0} \oplus \bar{a} = \overline{0 + a} = \bar{a}$$
and $\bar{a} \oplus \bar{0} = \overline{a + 0} = \bar{a},$

the element $\bar{0} \in \mathbb{Z}_n$ is an identity with respect to $\oplus$.

Axiom 4 Let $\bar{a} \in \mathbb{Z}_n$. Then $\overline{n - a} \in \mathbb{Z}_n$ and
$$\bar{a} \oplus \overline{n - a} = \overline{a + (n - a)} = \bar{n} = \bar{0}$$
Similarly, $\overline{n - a} \oplus \bar{a} = \overline{(n - a) + a} = \bar{n} = \bar{0}.$

Thus $\overline{n - a}$ is the inverse of $\bar{a}$ with respect to $\oplus$, and hence axiom 4 is satisfied. Thus $(\mathbb{Z}_n, \oplus)$ is a group. To show that it is *abelian*, we note that
$$\bar{a} \oplus \bar{b} = \overline{a + b} = \overline{b + a} = \bar{b} \oplus \bar{a}. \ \blacksquare$$

Example 6.1.5

The following table is the Cayley table for $(\mathbb{Z}_5, \oplus)$:

$\oplus$	$\bar{0}$	$\bar{1}$	$\bar{2}$	$\bar{3}$	$\bar{4}$
$\bar{0}$	$\bar{0}$	$\bar{1}$	$\bar{2}$	$\bar{3}$	$\bar{4}$
$\bar{1}$	$\bar{1}$	$\bar{2}$	$\bar{3}$	$\bar{4}$	$\bar{0}$
$\bar{2}$	$\bar{2}$	$\bar{3}$	$\bar{4}$	$\bar{0}$	$\bar{1}$
$\bar{3}$	$\bar{3}$	$\bar{4}$	$\bar{0}$	$\bar{1}$	$\bar{2}$
$\bar{4}$	$\bar{4}$	$\bar{0}$	$\bar{1}$	$\bar{2}$	$\bar{3}$

You should check directly from the table, as in example 6.1.2, that $(\mathbb{Z}_5, \oplus)$ satisfies axioms 1, 3 and 4. What property of the table shows that the group is abelian? (See problem 6.1.2a, 11.)

Multiplication in $\mathbb{Z}_n$

We can define a multiplicative operation $\otimes$ in $\mathbb{Z}_n$ by the rule

$$\bar{a} \otimes \bar{b} = \overline{ab}.$$

As with the additive operation $\oplus$, it is necessary to show that $\otimes$ is *well-defined*.

RESULT 6.1.3 The operation $\otimes$ on $\mathbb{Z}_n$ is well-defined.

Proof Let $\bar{a}, \bar{b} \in \mathbb{Z}_n$. By the definition of $\otimes$, we have

$$\bar{a} \otimes \bar{b} = \overline{ab}.$$

Now suppose that $a_1 \in \bar{a}$, $b_1 \in \bar{b}$. Then we could have chosen a_1, b_1 as class representatives of $\bar{a}$, $\bar{b}$ respectively. In this case we would have

$$\bar{a} \otimes \bar{b} = \bar{a}_1 \otimes \bar{b}_1 = \overline{a_1 b_1}.$$

Thus we must show that $\overline{a_1 b_1} = \overline{ab}$.

Now $\quad a_1 \in \bar{a} \Rightarrow a_1 \equiv a(\bmod n) \Rightarrow a_1 = a + kn$, for some $k \in \mathbb{Z}$.
and $\quad b_1 \in \bar{b} \Rightarrow b_1 \equiv b(\bmod n) \Rightarrow b_1 = b + hn$, for some $h \in \mathbb{Z}$.
Hence $\quad a_1 b_1 = ab + (ah + bk + khn)n$
$\quad \Rightarrow a_1 b_1 \equiv ab(\text{mob } n) \Rightarrow \overline{a_1 b_1} = \overline{ab}$.

This shows that $\otimes$ is well-defined. ∎

RESULT 6.1.4 The operation $\otimes$ is associative in $\mathbb{Z}_n$.

Proof We have to show that $(\bar{a} \otimes \bar{b}) \otimes \bar{c} = \bar{a} \otimes (\bar{b} \otimes \bar{c})$, $\forall \bar{a}, \bar{b}, \bar{c} \in \mathbb{Z}_n$.

Now $\quad (\bar{a} \otimes \bar{b}) \otimes \bar{c} = \overline{ab} \otimes \bar{c} = \overline{(ab)c}$
and $\quad \bar{a} \otimes (\bar{b} \otimes \bar{c}) = \bar{a} \otimes \overline{bc} = \overline{a(bc)}$.

But $(ab)c = a(bc)$, since multiplication is associative in $\mathbb{Z}$. Thus $\otimes$ is also associative. ∎

Despite the fact that the operation $\otimes$ is well defined and associative on $\mathbb{Z}_n$, the complete set $\mathbb{Z}_n = \{\bar{0}, \bar{1}, \ldots, \overline{n-1}\}$, $n \geq 2$ is *never* a group with respect to $\otimes$ (see problem 6.1.2a, 4, but you can probably anticipate the difficulty). However, it is true that for all $n \geq 2$, we can always find a *proper* subset of $\mathbb{Z}_n$ which is a group with respect to $\otimes$.

Example 6.1.6

Show that the subset $S = \{\bar{1}, \bar{3}, \bar{7}, \bar{9}\}$ of $\mathbb{Z}_{10}$ is a group with respect to $\otimes$.

Solution As for example 6.1.2, we start by constructing the Cayley table of S. Note that since S is a subset of $\mathbb{Z}_{10}$, we are multiplying mod 10.

$\otimes$	$\bar{1}$	$\bar{3}$	$\bar{7}$	$\bar{9}$
$\bar{1}$	$\bar{1}$	$\bar{3}$	$\bar{7}$	$\bar{9}$
$\bar{3}$	$\bar{3}$	$\bar{9}$	$\bar{1}$	$\bar{7}$
$\bar{7}$	$\bar{7}$	$\bar{1}$	$\bar{9}$	$\bar{3}$
$\bar{9}$	$\bar{9}$	$\bar{7}$	$\bar{3}$	$\bar{1}$

Axiom 1 All the entries in the table are elements of S. Hence S is *closed* with respect to $\otimes$.

Axiom 2 Operation $\otimes$ is associative (see result 6.1.4).

Axiom 3 From the table (or directly, from the definition of $\otimes$),

$$\bar{1} \otimes \bar{a} = \bar{a} \otimes \bar{1} = \bar{a}.$$

Hence S contains an identity element, satisfying axiom 3.

Axiom 4 From the table we see that

$$\bar{1} \otimes \bar{1} = \bar{1}; \qquad \bar{9} \otimes \bar{9} = \bar{1}; \qquad \bar{3} \otimes \bar{7} = \bar{7} \otimes \bar{3} = \bar{1}.$$

Hence every element of S has an inverse in S.
Thus $(S, \otimes)$ satisfies all four axioms and so is a group.

PROBLEMS 6.1.2a

1 Show that the set $K = \left\{ \begin{pmatrix} 1 & 0 \\ 0 & 1 \end{pmatrix}, \begin{pmatrix} -1 & 0 \\ 0 & 1 \end{pmatrix}, \begin{pmatrix} 1 & 0 \\ 0 & -1 \end{pmatrix}, \begin{pmatrix} -1 & 0 \\ 1 & -1 \end{pmatrix} \right\}$ is a group with respect to matrix multiplication.

2 Construct the Cayley tables for $(\mathbb{Z}_3, \oplus)$ and $(\mathbb{Z}_4, \oplus)$.
Find the inverse of $\bar{1}$ in (a) $(\mathbb{Z}_3, \oplus)$ and (b) $(\mathbb{Z}_4, \oplus)$.

3 Show that the operation $\otimes$ on the set $\mathbb{Z}_n$ admits an identity element, for all $n \geqslant 2$.

4 Construct the Cayley table for $(\mathbb{Z}_5, \otimes)$ and show that this is *not* a group.
Prove that $(\mathbb{Z}_n, \otimes)$ is not a group when $n \geqslant 2$.

5 The set $\mathbb{Z}_n^* = \{\bar{1}, \bar{2}, \ldots, \overline{n-1}\}$, $n \geqslant 2$. Decide, by constructing the Cayley table or otherwise, which of the following are groups with respect to $\otimes$.
(a) $\mathbb{Z}_3^*$;   (b) $\mathbb{Z}_4^*$;   (c) $\mathbb{Z}_5^*$;   (d) $\mathbb{Z}_6^*$;   (e) $\mathbb{Z}_2^*$.

6 Prove that $(\mathbb{Z}_n^*, \otimes)$ is not a group if n is a *composite* integer.

7 Prove that the subset $S = \{\bar{1}, \bar{3}, \bar{5}, \bar{7}\}$ of $\mathbb{Z}_8$ is a group with respect to $\otimes$.

8 Show that the subset $T = \{\bar{1}, \bar{3}, \bar{5}, \bar{7}\}$ of $\mathbb{Z}_{10}$ is *not* a group with respect to $\otimes$.

9 Let z_1, z_2 be complex numbers satisfying the equation

$$z^n = 1, \qquad n \in \mathbb{Z}^+,$$

that is, they are both complex nth roots of unity. Show that the following complex numbers also satisfy the equation $z^n = 1$:

$$\text{(a) } z = z_1 z_2; \qquad \text{(b) } z = \frac{1}{z_1}.$$

Hence, or otherwise, prove that the set of complex nth roots of unity is a group with respect to multiplication. Show that the group $(H, \times)$ of example 6.1.2 is a particular example of a group in this family.

10 Let X be the set $\mathbb{R} - \{0, 1\}$. Let f_i, $i = 1, 2 \ldots, 6$, be mappings of $X \to X$ defined by:

$$f_1(x) = x, \ f_2(x) = \frac{1}{x}, \ f_3(x) = 1 - x, \ f_4(x) = \frac{1}{1-x},$$

$$f_5(x) = \frac{x}{x-1}, \ f_6(x) = 1 - \frac{1}{x}.$$

Show that (a) $f_2 \circ f_3 = f_4$ and (b) $f_4 \circ f_6 = f_1$.
Construct a Cayley table for the set $M = \{f_1, f_2, \ldots f_6\}$ with respect to the operation composition of mappings, entering $f_i \circ f_j$ in the row opposite f_i and the column below f_j.
Hence show that $(M, \circ)$ is a group. Show that $(M, \circ)$ is non-abelian.

11 How could you tell from looking at the Cayley table of a group whether it is abelian?

PROBLEMS 6.1b

1 Decide which of the following sets are groups with respect to $\otimes$:

(a) the subset $\{\bar{3}, \bar{5}, \bar{7}\}$ of $\mathbb{Z}_8$; (b) the subset $\{\bar{1}, \bar{3}, \bar{5}, \bar{7}\}$ of $\mathbb{Z}_{12}$;
(c) the subset $\{\bar{1}, \bar{5}, \bar{7}, \bar{11}\}$ of $\mathbb{Z}_{12}$; (d) the subset $\{\bar{1}, \bar{5}\}$ of $\mathbb{Z}_8$;
(e) the subset $\{\bar{1}, \bar{5}\}$ of $\mathbb{Z}_{10}$; (f) the set $\mathbb{Z}_7^*$.

2 Prove that the subset $H = \{\bar{2}, \bar{4}, \bar{6}, \bar{8}\}$ of $\mathbb{Z}_{10}$ is a group with respect to $\otimes$. Give the identity element and the inverse of each element.

3 Find a subset of
(a) $\mathbb{Z}_{10}$ which forms a group of order 5 with respect to $\otimes$
(b) $\mathbb{Z}_{10}$ which forms a group of order 2 with respect to $\otimes$;
(c) $\mathbb{Z}_7$ which forms a group of order 3 with respect to $\otimes$.

4 Show that the set $\{2^m \, 5^n : m, n \in \mathbb{Z}\}$ is a group with respect to multiplication.

5 The set $5\mathbb{Z} = \{5m : m \in \mathbb{Z}\}$. Prove that $(5\mathbb{Z}, +)$ is a group. Is $(5\mathbb{Z}, \times)$ also a group?

6 Prove that the set of matrices $K = \left\{ \begin{pmatrix} x & -x \\ -x & x \end{pmatrix} : x \in \mathbb{R}^* \right\}$ is a group with respect to matrix multiplication, giving its identity and the inverse of a typical element.

7 Show that the set of rational numbers $\mathbb{Q}$ is not a group with respect to the operation $\circ$ defined by

$$a \circ b = a + b + 2ab.$$

Find the subset S of $\mathbb{Q}$ which is a group with respect to $\circ$, substantiating your answer.

8 Let $S = \{(x, y) : x \in \mathbb{R}, y \in \mathbb{R}^*\}$. An operation $\circ$ is defined on S by

$$(x_1, y_1) \circ (x_2, y_2) = (x_1 \, y_2 + y_1 \, x_2, \, y_1 \, y_2).$$

Prove that $(S, \circ)$ is a group.

9 Let a, b be given rationals, $a \neq 0$, and define the mapping $\alpha_{a, b} : \mathbb{R} \to \mathbb{R}$ by $\alpha_{a, b}(x) = ax + b$. Show that

(a) $\alpha_{a, b} \circ \alpha_{c, d} = \alpha_{p, q}$, where $p = ac$, $q = ad + b$ and $\circ$ denotes composition of mappings;

(b) $\alpha_{a, b}^{-1} = \alpha_{s, t}$, where $s = \dfrac{1}{a}$, $t = -\dfrac{b}{a}$.

Hence prove that the set $M = \{\alpha_{a, b} : a \in \mathbb{Q}^*, b \in \mathbb{Q}\}$ is a group with respect to composition of mappings.

6.2 DEDUCTIONS FROM THE AXIOMS

In section 6.1 we encountered some concrete examples of groups with various kinds of elements and different operations. However, you may have noticed that all these groups had some simple properties in common. For example, each group contained only one identity element; each group element had only one inverse; and in the case of a finite group, you probably noticed that each group element appeared exactly once in each row and column of the Cayley table. In this section, we show that these and some others, are properties of *all* groups, because they are results which can be deduced from the group axioms.

Let G be a group with respect to the operation $\circ$.

RESULT 6.2.1   The identity element of G is unique.

Proof   See result 5.3.1. ■

RESULT 6.2.2   Each element of G has a *unique* inverse in G.

Proof   Let $a \in G$. Then a has at least one inverse in G, by axiom 4. Suppose that a has more than one inverse in G, say b_1 and b_2.

Then $\quad a \circ b_1 = b_1 \circ a = e$ $\left.\right\}$ by definition of *inverse*.
and $\quad\ a \circ b_2 = b_2 \circ a = e$

Now consider the combination $a_1 \circ a \circ b_2$. There are two ways of inserting brackets in this expression and they both give the same result, since $\circ$ is *associative*.

Hence $(b_1 \circ a) \circ b_2 = b_2 \circ (a \circ b_2)$
$\quad\Rightarrow e \circ b_2 = b_1 \circ e$, by definition of *inverse*
$\quad\Rightarrow \quad\ b_2 = b_1$, by definition of e.

Hence every element of G has a unique inverse in G. ■

Notation

We usually denote the unique inverse of the element $a \in G$ by a^{-1}, unless we are given that the group operation is additive (such as addition of real or complex numbers, addition mod m, addition of matrices, etc.). The inverse of $a \in G$ with respect to an *additive* operation is often denoted by $(-a)$.

RESULT 6.2.3 The identity element e is its own inverse.

Proof $e \circ e = e$, since e is the identity
$\Rightarrow e$ is an inverse of e, by definition of *inverse*
$\Rightarrow e = e^{-1}$, since the inverse of e is unique. ∎

RESULT 6.2.4 Let $a \in G$. Then a is the inverse of a^{-1}.
That is: $\qquad (a^{-1})^{-1} = a$.

Proof $a \circ a^{-1} = a^{-1} \circ a = e$, by definition of *inverse*
$\Rightarrow a$ is an inverse of a^{-1},
$\Rightarrow a = (a^{-1})^{-1}$, since the inverse of a^{-1} is unique. ∎

RESULT 6.2.5 Let $a, b \in G$. Then $(a \circ b)^{-1} = b^{-1} \circ a^{-1}$.

Proof Since G is *closed* with respect to $\circ$, by axiom 1, $a \circ b$ is an element of G and hence $(a \circ b)^{-1} \in G$, by axiom 4.
Let $x = (a \circ b)^{-1}$.

Then $x \circ (a \circ b) = e$, by definition of *inverse*,
$\Rightarrow \quad (x \circ a) \circ b = e$, since $\circ$ is associative,
$\Rightarrow \quad \{(x \circ a) \circ b\} \circ b^{-1} = e \circ b^{-1}$
$\Rightarrow \quad (x \circ a) \circ (b \circ b^{-1}) = e \circ b^{-1}$, since $\circ$ is associative,
$\Rightarrow \quad (x \circ a) \circ e = e \circ b^{-1}$, by definition of *inverse*,
$\Rightarrow \quad x \circ a = b^{-1}$, by definition of e,
$\Rightarrow \quad (x \circ a) \circ a^{-1} = b^{-1} \circ a^{-1}$
$\Rightarrow \quad x \circ (a \circ a^{-1}) = b^{-1} \circ a^{-1}$, since $\circ$ is associative,
$\Rightarrow \quad x \circ e = b^{-1} \circ a^{-1}$, by definition of *inverse*,
$\Rightarrow \quad x = b^{-1} \circ a^{-1}$, by definition of e. ∎

This last result needs noting carefully, as the order of the elements is at first sight unexpected. We look at the combination $(a \circ b) \circ (b^{-1} \circ a^{-1})$

and show how by using the associative property of $\circ$ and the definition of the inverses and identity we get $(a \circ b) \circ (b^{-1} \circ a^{-1}) = e$.

$$(a \circ b) \circ (b^{-1} \circ a^{-1}) = a \circ (b \circ b^{-1}) \circ a^{-1} = a \circ e \circ a^{-1} = a \circ a^{-1} = e.$$

Similarly, we can show that $(b^{-1} \circ a^{-1}) \circ (a \circ b) = e$. You may be familiar with a special case of this result; the rule for finding the inverse of the product AB of two non−singular matrices A and B is given by $(AB)^{-1} = B^{-1}A^{-1}$.

THEOREM 6.2.6 *The cancellation laws*

    (a) $a \circ b = a \circ c \Rightarrow b = c$, $\forall\ a, b, c \in G$.

    (b) $b \circ a = c \circ a \Rightarrow b = c$, $\forall\ a, b, c \in G$.

Proof Since $a \in G$, a has an inverse $a^{-1} \in$ G.

(a) Then $a \circ b = a \circ c$

    $\Rightarrow$ $a^{-1} \circ (a \circ b) = a^{-1} \circ (a \circ c)$

    $\Rightarrow$ $(a^{-1} \circ a) \circ b = (a^{-1} \circ a) \circ c$, since $\circ$ is associative,

    $\Rightarrow$ $e \circ b = e \circ c$, by definition of inverse,

    $\Rightarrow$ $b = c$, by definition of e.

(b) Similarly, $b \circ a = c \circ a \Rightarrow (b \circ a) \circ a^{-1} = (c \circ a) \circ a^{-1}$. The completion of the proof is left to the reader. ∎

THEOREM 6.2.7 Let $a, b \in G$. Then the equations (a) $a \circ x = b$, (b) $y \circ a = b$ have the unique solutions in G given by $x = a^{-1} \circ b$ and $y = b \circ a^{-1}$, respectively.

Proof Since $a \in G$, a has an inverse $a^{-1} \in G$, by axiom 4, and hence since $b \in G$, the elements $x = a^{-1} \circ b$, $y = b \circ a^{-1} \in G$ by axiom 1.

(a) Then $a \circ x = a \circ (a^{-1} \circ b) = (a \circ a^{-1}) \circ b = e \circ b = b$.
Hence $x = a^{-1} \circ b$ is a solution of $a \circ x = b$.
Suppose $x = x_1$ and $x = x_2$ are two solutions of $a \circ x = b$, where $x_1, x_2 \in G$. Then $b = a \circ x_1 = a \circ x_2$.
But $a \circ x_1 = a \circ x_2 \Rightarrow x_1 = x_2$, by theorem 6.2.6.
Hence the solution $x = a^{-1} \circ b$ is the *unique* solution in G of the equation $a \circ x = b$.

(b) The proof that $y = b \circ a^{-1}$ is the unique solution of the equation $y = b \circ a^{-1}$ is left to the reader. ∎

This last result can be used to establish a feature of the Cayley table of a finite group that we noted at the beginning of the section.

RESULT 6.2.8   Let G be a finite group. Then every element of G occurs exactly once in each row (column) of the Cayley table of G.

Proof Let $G = \{g_1, g_2, \ldots, g_n\}$ and let g denote a typical element of G. Consider the row of the Cayley table opposite g; it consists of the elements $g \circ g_1, g \circ g_2, \ldots, g \circ g_n$. But given $g_i \in G$, we can find a unique element $x \in G$ such that $g \circ x = g_1$, by theorem 6.2.7 (a). Hence the element g_i occurs in this row exactly once, in the column below x.

A very similar argument holds for the columns of the Cayley table, using theorem 6.2.7 (b). ∎

Note 1: The results we have proved in this section are true for *all* groups because they have been deduced directly from the group axioms. In order to show that a set S is *not* a group with respect to a given operation $\circ$, it is sufficient to find just *one* axiom, or just *one* of the results proved in this section, that is *not* satisfied by $(S, \circ)$.

Note 2: Great care must be taken never to alter the *order* in which the group elements occur in any combination of two or more elements, unless we are told that the group is abelian. On the other hand, the group operation is always associative, so we may rebracket or insert brackets into a combination of three or more elements in any way we please.

PROBLEMS   6.2a

1 Let $S = \{a, b, c\}$. The following tables define binary operations $\circ$ and $*$ on S. Explain why neither $(S, \circ)$ nor $(S, *)$ is a group.

(a)

$\circ$	a	b	c
a	a	b	c
b	b	a	b
c	c	b	a

(b)

$*$	a	b	c
a	b	a	c
b	a	c	b
c	c	b	a

2 Suppose that $S = \{x, y\}$ is a group with respect to a binary operation $\circ$. Show that there is only one way to fill in the Cayley table for S if we are given that x is the identity.

3 (a) Let $M(2, \mathbb{R})$ be the set of all 2×2 matrices over $\mathbb{R}$. Give an example of matrices $A, B, C \in M(2, \mathbb{R})$ such that $AB = AC$, but $B \neq C$. What type of matrix has to be chosen for A? Explain why. Is $M(2, \mathbb{R})$ a group with respect to matrix multiplication?

(b) Let $GL(2, \mathbb{R})$ be the subset of all non-singular matrices in $M(2, \mathbb{R})$. Find an example of matrices $A, B, C \in GL(2, \mathbb{R})$ such that $AB = CA$, but $B \neq C$. Explain why this does not violate the cancellation law for a group.

4 Let $(G, \circ)$ be a group and let a be a fixed element of G. Define a mapping $\alpha: G \to G$ by $\alpha(x) = a \circ x$, for every $x \in G$. Prove that α is bijective. Define the inverse mapping α^{-1} of α.

5 Let $(G, \circ)$ be a group. Denote $x \circ x$ by x^2, $x \in G$. Prove that G is abelian if and only if $(a \circ b)^2 = a^2 \circ b^2$, for every $a, b \in G$.

6 Let $(G, \circ)$ be a group with identity element e and let $a, b, c \in G$, be such that $a \circ b \circ c = e$. Show that $b \circ c \circ a = e$. Which of the following combinations can be proved to give the identity: $a \circ c \circ b$, $b \circ a \circ c$, $c \circ a \circ b$, $c \circ b \circ a$?

PROBLEMS 6.2b

1 The following is part of a Cayley table for a group $G = \{a, b, c, d\}$, with respect to an operation $\circ$.

$\circ$	a	b	c	d
a	c	.	.	.
b	.	.	.	a
c	.	.	.	d
d	.	.	.	.

(a) Which is the identity element of $(G, \circ)$?

(b) Show that there is only one way to fill in the rest of the table, justifying each step.

Note: You are not required to verify that $\circ$ is associative.

2 Complete the proof of theorem 6.2.6.

3 Complete the proof of theorem 6.2.7.

4 Let $(G, \circ)$ be a finite group. Prove that every element of G occurs exactly once in each column of the Cayley table of G.

5 Let $(G, \circ)$ be a group. Prove that G is abelian if and only if $(x \circ y)^{-1} = x^{-1} \circ y^{-1}$, for every $x, y \in G$.

6 Let $(G \circ)$ be a group with identity element e. Suppose that $a, b, c, d \in G$ are such that $a \circ b \circ c \circ d = e$. Prove that $b \circ c \circ d \circ a = e$.

6.3 POWERS OF AN ELEMENT

Notation

In dealing with an abstract group $(G, \circ)$, it is common practice to simplify the notation and write ab to mean $a \circ b$, for any $a, b \in G$, provided no confusion can arise. We then refer to the group $(G, \circ)$ simply as 'the group G', with no specific mention of the operation.

In this notation, axiom 2 reads:

$$(ab)c = a(bc), \quad \forall \, a, b, c \in G,$$

for example.

From now on, we will always use e to denote the identity element of an abstract group G.

Powers of an element

Let a be any element of a group G. Since the group operation is always *binary* and *associative*, it follows that combinations such as aa, aaa, $aaaa$ etc. each have a unique meaning in G and hence by a natural extension of the notation above, we denote them by a^2, a^3, a^4 etc. In general, we define the element a^n by:

$$a^0 = e;$$
$$a^n = aaa \ldots a \ (n \text{ factors}), \ \forall \, n \in \mathbb{Z}^+;$$
$$a^{-n} = (a^{-1})^n, \ \forall \, n \in \mathbb{Z}^+.$$

From these definitions, it is easy to show that the familiar laws of indices satisfied by real numbers are true in a group when the indices are integers.

RESULT 6.3.1   Let G be a group and let $a \in G$. Then for any $m, n \in \mathbb{Z}$, we have

(a) $a^m a^n = a^{m+n}$,
(b) $(a^m)^n = a^{mn}$,
(c) $(a^n)^{-1} = a^{-n}$.

Additive notation

We have a convention that addition is only used as the group operation when the group is *abelian*. In additive notation, a^2 becomes $a + a$, a^3 becomes $a + a + a$, etc., and so we usually write na in place of a^n. For an additive group $(G, +)$, result 6.3.1 becomes:

(a) $ma + na = (m + n)a$
(b) $n(ma) = (nm)a$
(c) $-(na) = (-n)a$.

Order of an element

Let G be a *finite* group. Let $a \in G$ and consider the infinite sequence of powers of a:

$$\ldots, a^{-2}, a^{-1}, a^0 = e, a, a^2, a^3, a^4, \ldots$$

Since G is closed, all powers of a are elements of G. But since G contains only a finite number of distinct elements, this sequence must contain repetitions. Hence we can find integers s, t such that

$$a^t = a^s, \quad \text{where } s < t.$$

But $\qquad\qquad (a^{-1})^s = a^{-s} \in G$

and hence $\qquad\quad a^t a^{-s} = a^s a^{-s}$

$$\Rightarrow a^{t-s} = a^0 = e, \quad \text{by result 6.3.1.}$$

Thus given any $a \in G$, where G is a *finite* group, we can find a positive integer q such that $a^q = e$.

If G is an *infinite* group it is not necessarily true that the sequence of powers of an element of $a \in G$ must contain repetitions. However, for some particular elements $a \in G$, we may be able to find an integer q such that $a^q = e$.

Examples 6.3.1

1. Let G be any group, finite or infinite, with identity element e. Then $e^q = e$, $\forall\ q \in \mathbb{Z}$.

2. Consider the infinite group $(\mathbb{C}^*, \times)$. The identity element $e = 1 + 0i$. Let $a = -1 + 0i$, $b = 0 + i$.

Then (a) $e = a^2 = a^4 = a^6 = \ldots$
   (b) $e = b^4 = b^8 = b^{12} = \ldots$

Let G be any group. Given any element $a \in G$, if we can find an integer q such that $a^q = e$, we say that a has *finite order*. In this case, the *order* of a is defined to be the *least* positive integer r such that $a^r = e$. We denote the order of a symbolically by $o(a)$. If no integer exists such that $a^q = e$, we say that a has *infinite* order. We write this as $o(a) = \infty$.

Examples 6.3.2

1. In $(\mathbb{C}^*, \times)$, the element $-1 + 0i$ has order 2 and $0 + i$ has order 4.

2. Consider the set $\mathbb{Z}_7^* = \{\bar{1}, \bar{2}, \bar{3}, \bar{4}, \bar{5}, \bar{6}\}$ with respect to $\otimes$ (multiplication mod 7). It is easy to show by making a Cayley table, or otherwise, that this is a group. Clearly the identity element is $\bar{1}$.
   We will find the order of the elements $\bar{2}$ and $\bar{3}$.

   (a) $a = \bar{2}$; $a^2 = \bar{2} \otimes \bar{2} = \bar{4}$; $a^3 = a^2a = \bar{4} \otimes \bar{2} = \bar{1}$
   and hence $o(\bar{2}) = 3$.
   (b) $a = \bar{3}$; $a^2 = \bar{3} \otimes \bar{3} = \bar{2}$; $a^3 = a^2a = \bar{2} \otimes \bar{3} = \bar{6}$;
   $a^4 = a^3a = \bar{6} \otimes \bar{3} = \bar{4}$; $a^5 = a^4a = \bar{4} \otimes \bar{3} = \bar{5}$;
   $a^6 = a^5a = \bar{5} \otimes \bar{3} = \bar{1}$, and hence $o(\bar{3}) = 6$.

3. We know from section 6.1 that $\mathbb{Z}_8$ is a group with respect to $\oplus$ (addition mod 8), with identity element $\bar{0}$.
   We will find the order of the elements $\bar{5}$ and $\bar{6}$. Notice that since the group operation is *additive*, we will be looking at the sequence of *sums*, $a \oplus a$, $a \oplus a \oplus a$, etc. or, in other words, at the multiples $2a$, $3a$, $4a, \ldots$, until we find the least value of $r > 0$ such that $ra = \bar{0}$.

   (a) $a = \bar{5}$; $2a = \bar{2}$; $3a = \bar{7}$; $4a = \bar{4}$; $5a = \bar{1}$;
      $6a = \bar{6}$; $7a = \bar{3}$; $8a = \bar{0}$. Hence $o(\bar{5}) = 8$.
   (b) $a = \bar{6}$; $2a = \bar{4}$; $3a = \bar{2}$; $4a = \bar{0}$. Hence $o(\bar{6}) = 4$.

159

RESULT 6.3.2   Let G be a group.

(a) $o(e) = 1$;
(b) $o(a^{-1}) = o(a)$, $\forall a \in G$.

Proof (a) Since $e^n = e$, $\forall$, $n \in \mathbb{Z}^+$, $o(e) = 1$ follows from the definition of *order*.

(b) We note that for any $m \in \mathbb{Z}$,

$$a^m(a^{-1})^m = a^m a^{-m}, \text{ by definition,}$$
$$= e, \text{ by result 6.3.1(c).}$$

Hence $a^m = e \Leftrightarrow (a^{-1})^m = e$, $\forall\, m \in \mathbb{Z}$.

Thus the least positive value of m for which $a^m = e$ is also the least positive value of m for which $(a^{-1})^m = e$. *Hence a and a^{-1} either both have the same finite order or* $a^m \neq e$ for any $m \in \mathbb{Z}$ and $o(a) = o(a^{-1}) = \infty$. ∎

RESULT 6.3.3.   Let G be a group and suppose $a \in G$ is an element of order m. Then

(a) $a^s = e \Leftrightarrow m|s$;
(b) $a^t = a^s \Leftrightarrow t \equiv s \pmod{m}$.

Proof (a) Suppose first $m|s$. Then we can find an integer q such that $s = mq$.

Hence $a^s = a^{mq} = (a^m)^q = e^q = e$.

Conversely, suppose $a^s = e$. Then by the division theorem we can write $s = mq + r$, where $q, r \in \mathbb{Z}$ and $0 \leqslant r < m$.

Then $a^s = a^{mq+r} = a^{mq}a^r = ea^r = a^r$.

Hence $a^s = e \Rightarrow a^r = e \Rightarrow r = 0$ (from the definition of m)
$$\Rightarrow m|s.$$

(b) If $a^t = a^s$, then $a^{t-s} = e$ and hence $m|t-s$ by (a). Thus $a^t = a^s \Rightarrow t \equiv s \pmod{m}$. The converse is proved by reversing the argument. ∎

PROBLEMS 6.3a

1 Find the order of each of the elements $\bar{4}, \bar{5}, \bar{6}$ in the group $(\mathbb{Z}_7^*, \otimes)$ (see example 6.3.2, 2). Show that every element of this group can be written as a power of the element $\bar{5}$. Which elements is it not possible to write as a power of $\bar{2}$?

2 Find the order of each of the elements $\bar{2}, \bar{3}, \bar{4}$ in the group $(\mathbb{Z}_8, \oplus)$ (see example 6.3.2, 3). Show that every element of this group can be written as a multiple of the element $\bar{3}$.

3 Find the order of each of the elements of the group $S = \{\bar{1}, \bar{3}, \bar{7}, \bar{9}\}$ with respect to multiplication mod 10 (see example 6.1.6).

4 Let $z = r (\cos \theta + i \sin \theta)$ be an element of $(\mathbb{C}^*, \times)$. Prove that z has finite order if and only if $r = 1$ and $\theta = \dfrac{2k\pi}{m}$ for some integers m and k.

5 Rewrite the following results in *additive* group notation:
(a) $a^t = a^s \Leftrightarrow t \equiv s \pmod{m}$;   (b) $(a^{-1})^{-1} = a$;
(c) $ab = ac \Leftrightarrow b = c$.

6 (a) Explain why the result $(ab) = b^{-1}a^{-1}$ can be written as $-(a + b) = (-a) + (-b)$ in *additive* group notation.
(b) Let $(G, +)$ be an additive group and suppose $a, b \in G$. Give the unique solutions $x, y \in G$ to the equations

$$a + x = b; \qquad y + a = b.$$

Does $x = y$?

7 Let G be a group and let $x, y \in G$. Write down what is meant by the element $(xy)^r$. (Be careful not to assume that $xy = yx$.) Show that $(xy)^r = y^{-1}(yx)^r y$ and hence prove that $o(xy) = o(yx)$.

PROBLEMS 6.3b

1 Find the order of each of the following matrices in the group GL $(2, \mathbb{R})$, the multiplicative group of 2×2 non-singular matrices over $\mathbb{R}$:

(a) $a = \begin{bmatrix} 1 & -1 \\ 0 & -1 \end{bmatrix}$; (b) $b = \begin{bmatrix} -1 & 0 \\ 0 & -1 \end{bmatrix}$; (c) $c = \begin{bmatrix} 1 & -1 \\ 0 & -1 \end{bmatrix}$.

2 Find the order of each of the mappings f_i in the group $(M, \circ)$ defined in problem 6.1.2a, 10.

3 Let G be a group and $a, x \in G$. Suppose $b = x^{-1}ax$.
Prove by induction that $b^n = x^{-1}a^nx$, $\forall\ n \in \mathbb{Z}^+$.
Hence prove that $o(a) = o(b)$.

4 Let G be a group and $x \in G$. Suppose $o(x) = n$ and m is an integer
such that $\gcd(m, n) = d$. Prove that $o(x^m) = n/d$.

6.4 CYCLIC GROUPS

Let G be a group. Suppose we can find an element $a \in G$ such that

$$G = \{a^r : r \in \mathbb{Z}\}.$$

Then G is said to be a *cyclic* group *generated by* the element a. The
element a is called a *generator* of G.

Notation

If G is a cyclic group generated by an element a, we write

$$G = \langle a \rangle.$$

Examples 6.4.1

1. Let $G = \{ \ldots, \frac{1}{8}, \frac{1}{4}, \frac{1}{2}, 1, 2, 4, 8, \ldots \}$. It is easy to verify that G is a
   group with respect to multiplication with identity element 1. Further,
   $G = \{2^r : r \in \mathbb{Z}\}$ and hence G is cyclic, generated by the element 2.

2. We have shown in example 6.1.6 that the set $S = \{\bar{1}, \bar{3}, \bar{7}, \bar{9}\}$ is a group
   with respect to multiplication mod 10. The table here shows that
   $S = \langle \bar{3} \rangle$:

r	0	1	2	3	4 $\ldots$
$(\bar{3})^r$	$\bar{1}$	$\bar{3}$	$\bar{9}$	$\bar{7}$	$\bar{1} \ldots$

Cyclic groups in additive notation

Suppose that G is a group with respect to an additive operation. If we
can find an element $a \in G$ such that

$$G = \{ra: r \in \mathbb{Z}\},$$

then G is cyclic, generated by the element a. Again, we write $G = \langle a \rangle$.

Examples 6.4.2

1. Consider the group $(\mathbb{Z}, \oplus)$. Clearly $\mathbb{Z} = \{r(1): r \in \mathbb{Z}\}$ and hence $(\mathbb{Z}, \oplus)$ is cyclic, generated by the element 1.

2. We know that $(\mathbb{Z}_5, \oplus)$ is a group. The following table shows that $(\mathbb{Z}_5, \oplus) = \langle \bar{1} \rangle$:

r	0	1	2	3	4	5	...
$r(\bar{1})$	$\bar{0}$	$\bar{1}$	$\bar{2}$	$\bar{3}$	$\bar{4}$	$\bar{0}$	...

The following results show the connection between the *order* of a finite cyclic group and the *order* of a generator.

RESULT 6.4.1   Let G be a finite cyclic group generated by a. Then

$$o(a) = |G|.$$

Proof   Let $|G| = n$ and suppose that $o(a) = m$. We want to show $m = n$. Consider the set $S = \{e, a, a^2, \dots, a^{m-1}\}$. By result 6.3.3, the elements of S are distinct and hence $|S| = m$. Clearly $S \subseteq G$.

Let $x \in G$. Then since $G = \langle a \rangle$, $x = a^r$, for some $r \in \mathbb{Z}$. But by result 6.3.3, $a^r = a^s$, where $0 \leq s \leq m-1$, and hence $x \in S$. Thus $G \subseteq S$ and this taken with $S \subseteq G$ shows $S = G$ and $m = n$. ∎

RESULT 6.4.2   Let G be a finite group of order n. Then G is cyclic if and only if G contains an element a of order n.

Proof   (i) Suppose G is cyclic. Then we can find a generator a of G. But $o(a) = |G| = n$, by result 6.4.1. Hence G contains an element a of order n.

(ii) Now suppose G contains an element a of order n. Then, as in the proof of result 6.4.1, the elements of the set $S = \{e, a, a^2, \ldots, a^{n-1}\}$ are all distinct and hence $|S| = n$. But $S \subseteq G$ and hence $S = G$. Thus $G = \langle a \rangle$ and so is cyclic. ∎

Examples 6.4.3

1. Show that the group $G = \{ \ldots, \frac{1}{8}, \frac{1}{4}, \frac{1}{2}, 1, 2, 4, 8, \ldots \}$ of example 6.4.1, 1 has more than one generator.

   Solution We have noted that $G = \{2^r : r \in \mathbb{Z}\}$ and hence $G = \langle 2 \rangle$.
   Consider the set $G_1 = \{(\frac{1}{2})^r : r \in \mathbb{Z}\}$.
   If $x \in G_1$, then $x = (\frac{1}{2})^r$, for some $r \in \mathbb{Z}$. But $(\frac{1}{2})^r = 2^{-r} \in G$.
   Similarly, if $y \in G$, then $y = 2^s = (\frac{1}{2})^{-s} \in G_1$.
   Hence $G_1 \subseteq G$ and $G \subseteq G_1$, so $G = G_1 = \langle \frac{1}{2} \rangle$.

2. Find all the generators of the group $S = \{\bar{1}, \bar{3}, \bar{7}, \bar{9}\}$ of example 6.4.1, 2.

   Solution We have already shown that S is cyclic and that $S = \langle \bar{3} \rangle$.
   Of the remaining elements, $o(\bar{1}) = 1$, $o(\bar{9}) = 2$ and $o(\bar{7}) = 4$.
   Hence by result 6.4.1, $\bar{1}$ and $\bar{9}$ are *not* generators of S but $\bar{7}$ is a generator. Hence the only generators of S are $\bar{3}$ and $\bar{7}$.

3. Find all the generators of the group $(\mathbb{Z}_5, \oplus)$.

   Solution We have already shown that $(\mathbb{Z}_5, \oplus) = \langle \bar{1} \rangle$ (see example 6.4.2, 2). $\bar{0}$ is clearly not a generator. However it is easy to verify that $o(\bar{2}) = o(\bar{3}) = o(\bar{4}) = 5$. Hence by result 6.4.1, the generators of $(\mathbb{Z}_5, \oplus)$ are $\bar{1}, \bar{2}, \bar{3}$ and $\bar{4}$.

It is clear from these examples that a cyclic group may have more than one generator. In fact it is not difficult to show that every infinite cyclic group has two generators (see problems 6.4a, 8 and 6.4b, 9) and that all but the very smallest cyclic groups (those of orders 1 and 2) have a least two generators and often more. The following result shows that once we have found one generator of a finite cyclic group, there is an easy method for finding the other generators.

RESULT 6.4.3   Let $G = \langle a \rangle$ be a finite cyclic group of order n. The generators of G are the elements a^r, where $\gcd(r, n) = 1$.

Proof
  (i) From result 6.4.1, we know that a^r is not a generator unless $o(a^r) = n$.
      Suppose that $\gcd(r, n) = d$. Let $n = dn_1$ and $r = dr_1$.
      Then $(a^r)^{n_1} = a^{rn_1} = a^{rn/d} = a^{r_1 n} = (a^n)^{r_1} = e$.
      Hence a^r has order n_1 (or a factor of n_1) by result 6.3.3.
      But if $d > 1$, then $n_1 < n$ and hence a^r is *not* a generator.
  (ii) Now suppose $\gcd(r, n) = 1$. Let $o(a^r) = m$. Then $a^{rm} = e$.
      But $o(a) = n$ and hence $n|rm$, by result 6.3.3.
      But $n|rm \Rightarrow n|m$, since $\gcd(r, n) = 1$, and hence $m \geq n$. Now $m > n$ is impossible because the powers $(a^r)^0$, $(a^r)^1$, $(a^r)^2, \ldots, (a^r)^{m-1}$ are distinct elements of G. Thus $m = n$ and hence a^r is a generator. ∎

Example 6.4.4

Consider again the group $S = \{\bar{1}, \bar{3}, \bar{7}, \bar{9}\}$ of example 6.4.1, 2. From the table, $S = \langle \bar{3} \rangle$ and further

$$\bar{1} = (\bar{3})^0 = (\bar{3})^4; \quad \bar{7} = (\bar{3})^3; \quad \bar{9} = (\bar{3})^2 .$$

In this case $n = |S| = 4$. Result 6.4.3 tells us that $(\bar{3})^r$ is a generator if and only if $\gcd(r, n) = 1$.
Since $\gcd(0, 4) \neq 1$, $\gcd(2, 4) \neq 1$, the elements $\bar{1}$ and $\bar{9}$ are not generators. But $\gcd(3, 4) = 1$ and hence $(\bar{3})^3 = \bar{7}$ is a generator.
Compare this method with example 6.4.3, 2.

You may have noticed that these examples of cyclic groups are all abelian. In fact *all* cyclic groups are abelian, as we show below.

RESULT 6.4.4   Let G be a cyclic group. Then G is abelian.

Proof   Since G is cyclic we can find an element $a \in G$ such that $G = \langle a \rangle$. Hence if $x, y \in G$, we can find integers r, s such that $x = a^r$ and $y = a^s$. Then $xy = a^r a^s = a^{r+s} = a^{s+r} = a^s a^r = yx$.
Hence every pair of elements in G commute and G is abelian. ∎

It is *not* true that every abelian group is cyclic. We give a counter-example below and others in problems 6.4a and 6.4b.

Example 6.4.5

The group $(\mathbb{Q}, +)$ is not cyclic.

Solution Suppose, if possible, we could find a generator $q \in \mathbb{Q}$.
Then $\mathbb{Q} = \{rq: r \in \mathbb{Z}\}$.
But $q/2 \in \mathbb{Q}$ and $q/2 \neq rq$ for any $r \in \mathbb{Z}$. This contradiction shows that $(\mathbb{Q}, +)$ is not cyclic.

PROBLEMS 6.4a

1 Show that the group $(H, \times)$ of example 6.1.2 is cyclic. Show that two of its elements are generators and two are not.

2 Show that the group $(K, \times)$, of problem 6.1.2a, 1, is *not* cyclic.

3 Give a reason why the group $(M, \circ)$, of problem 6.1.2a, 10, cannot be cyclic.

4 For each of the following groups, either explain why it is not cyclic or give a table to show how it can be written as $\langle g \rangle$ for a suitable g (see example 6.4.1, 2).

(a) $(S, \otimes)$, where S is the subset $\{\bar{1}, \bar{2}, \bar{4}, \bar{5}, \bar{7}, \bar{8}\}$ of $\mathbb{Z}_9$;
(b) $(T, \otimes)$, where T is the subset $\{\bar{1}, \bar{5}, \bar{7}, \overline{11}\}$ of $\mathbb{Z}_{12}$;
(c) $(\mathbb{Z}_7^*, \otimes)$.

5 Find all the elements of $(\mathbb{Z}_{12}), \oplus)$ of order (a) 2; (b) 3; (c) 4; (d) 6; (e) 12.
Explain why $(\mathbb{Z}_{12}, \oplus)$ cannot contain any element of order 5.

6 Find all the generators of (a) $(\mathbb{Z}_{10}, \oplus)$; (b) $(\mathbb{Z}_7, \oplus)$.

7 Prove that the group $M(2, \mathbb{Z})$ of all 2×2 matrices with integer entries under the operation matrix addition is *not* cyclic.

8 Suppose $G = \langle g \rangle$ is a cyclic group. Prove that g^{-1} is also a generator of G.

9 Prove that $(\mathbb{Q}^*, \times)$ is *not* cyclic.

10 Prove that the group of complex nth roots of unity (see problem 6.1.2a, 9) is cyclic.

PROBLEMS 6.4b

1 Prove that the group $(H, \otimes)$, where H is the subset $\{\bar{2}, \bar{4}, \bar{6}, \bar{8}\}$ of $\mathbb{Z}_{10}$ is cyclic. Find all the generators.

2 Show that the group $(S, \otimes)$, where S is the subset $\{\bar{1}, \bar{3}, \bar{5}, \bar{7}\}$ of $\mathbb{Z}_8$ is not cyclic.

3 Prove that the group $(\mathbb{Z}_{11}^*, \otimes)$ is cyclic. Apply results 6.4.3 to the first generator you find to give all the generators.

4 Find all the elements of $(\mathbb{Z}_{14}, \oplus)$ of order (a) 2; (b) 7; (c) 14.

5 Find all the generators of $(\mathbb{Z}_{20}, \oplus)$.

6 Prove that $(\mathbb{Z}_p, \oplus)$, where p is prime, has $p-1$ generators.

7 Prove that the group $(3\mathbb{Z}, \oplus)$ is cyclic and give two possible generators. Does this group have any other generators?

8 Let $G = \{2^m 3^n : m, n \in \mathbb{Z}\}$. Decide whether or not the group $(G, \times)$ is cyclic, justifying your answer.

9 Suppose that $G = \langle g \rangle$ is an *infinite* cyclic group. Prove that g and g^{-1} are the *only* generators of G.

10 Prove that if $G = \langle a \rangle$ is a finite cyclic group of order n then the order of every element of G is a divisor of n.

6.5 PERMUTATIONS AND THE SYMMETRIC GROUP

A *permutation* of a non-empty set X is a *bijective* mapping of X into X; we could describe it as a one–one correspondence between the elements of X.

Let $S(X)$ denote the set of all permutations on a given non-empty set X. We shall now collect together some of the results on mappings, established in previous chapters, to prove that $S(X)$ is a group with respect to composition of mappings.

Let α, β be typical permutations in $S(X)$. Then since Im $\alpha = X =$ Dom β, the composition mapping $\beta \circ \alpha$ is defined and hence $\circ$ is a *binary* operation on $S(X)$.

Axiom 1 Since $\alpha: X \to X$ and $\beta: X \to X$, $\beta \circ \alpha$ is also a mapping of X into X; also, α, β bijective $\Rightarrow \beta \circ \alpha$ is bijective (theorem 3.3.1). Hence $\alpha, \beta \in S(X) \Rightarrow \beta \circ \alpha \in S(X)$ and thus $S(X)$ is closed under $\circ$.

Axiom 2 Composition of mappings is associative (problem 5.5.2a, 6).

Axiom 3   Let ι denote the identity mapping on X. Then ι is bijective (result 3.4.1) and hence $\iota \in S(X)$. Also ι has the property that $\iota \circ \alpha = \alpha \circ \iota = \alpha$, $\forall \, \alpha \in S(X)$ (result 3.4.2).

Thus $S(X)$ contains an identity element with respect to $\circ$.

Axiom 4   Since α is bijective, α is invertible (theorem 3.4.3) and α^{-1} is also a bijective mapping of X into X (result 3.4.5). Further, $\alpha \circ \alpha^{-1} = \alpha^{-1} \circ \alpha = \iota$. Hence every element of $S(X)$ has an inverse in $S(X)$.

We have thus established the following:

THEOREM 6.5.1   The set $S(X)$ of permutations on a non-empty set X is a group with respect to composition of mappings.

We now consider the group $S(X)$ in the case when X is a finite set. Since we are interested only in the *permutations* of X, we may denote the elements of X by any symbols we choose. The simplest choice is to let $X = \{1, 2, \ldots, n\}$, where $|X| = n$.

The symmetric group, S_n

When $|X| = n$, the group $S(X)$ is known as the *symmetric group on n symbols* and denoted by S_n. In writing down the elements of S_n (i.e. the permutations of X), we assume that we have denoted the elements of X by the positive integers $1, 2, \ldots, n$.

Writing permutations

Let $\alpha \in S_n$. Then since $\alpha: X \to X$ is a mapping, it would be natural to specify α by giving a table, such as:

x	1	2	$\ldots$	n
$\alpha(x)$	$\alpha(1)$	$\alpha(2)$	$\ldots$	$\alpha(n)$

We simplify this notation by writing:

$$\alpha = \begin{pmatrix} 1 & 2 & \ldots & n \\ \alpha(1) & \alpha(2) & \ldots & \alpha(n) \end{pmatrix}.$$

Examples 6.5.1

1. Let $n = 5$. Then $\alpha = \begin{pmatrix} 1 & 2 & 3 & 4 & 5 \\ 5 & 3 & 1 & 4 & 2 \end{pmatrix}$ denotes the permutation in S_5

   defined by α: $\quad 1 \mapsto 5, \quad 2 \mapsto 3, \quad 3 \mapsto 1, \quad 4 \mapsto 4, \quad 5 \mapsto 2$.

2. Let $\beta \in S_4$ be defined by β: $1 \mapsto 3, \quad 3 \mapsto 2, \quad 2 \mapsto 4, \quad 4 \mapsto 1$.

   Then $\beta = \begin{pmatrix} 1 & 3 & 2 & 4 \\ 3 & 2 & 4 & 1 \end{pmatrix} = \begin{pmatrix} 1 & 2 & 3 & 4 \\ 3 & 4 & 2 & 1 \end{pmatrix}$.

   Notice that the *order* in which the columns are written does not matter. It is often useful to rearrange the columns, if necessary, so that the elements in the top row are in numerical order.

3. Let $\alpha = \begin{pmatrix} 1 & 2 & 3 & 4 \\ 2 & 3 & 1 & 4 \end{pmatrix}$, $\beta = \begin{pmatrix} 1 & 2 & 3 \\ 2 & 3 & 1 \end{pmatrix}$. Does $\alpha = \beta$?

   From the way the permutations are written, we see that

   Dom α = Im α = $\{1, 2, 3, 4\}$. While Dom β = Im β = $\{1, 2, 3\}$.

   Hence $\alpha \neq \beta$. In terms of the symmetric groups, $\alpha \in S_4$, while $\beta \in S_3$.

4. Let $\iota \in S_n$ denote the *identity permutation*. Then

   $$\iota = \begin{pmatrix} 1 & 2 & \ldots & n \\ 1 & 2 & \ldots & n \end{pmatrix}.$$

Composition of permutations

Suppose that $\alpha, \beta \in S$.

Let $\beta = \begin{pmatrix} 1 & 2 & \ldots & n \\ b_1 & b_2 & \ldots & b_n \end{pmatrix}$ and $\alpha = \begin{pmatrix} b_1 & b_2 & \ldots & b_n \\ a_1 & a_2 & \ldots & a_n \end{pmatrix}$.

Following our previous definition of the composition of two mappings, we define $\alpha \circ \beta$ by $(\alpha \circ \beta)(x) = \alpha(\beta(x))$, $x = 1, 2, \ldots, n$.

Hence $\begin{pmatrix} b_1 & b_2 & \ldots & b_n \\ a_1 & a_2 & \ldots & a_n \end{pmatrix} \circ \begin{pmatrix} 1 & 2 & \ldots & n \\ b_1 & b_2 & \ldots & b_n \end{pmatrix} = \begin{pmatrix} 1 & 2 & \ldots & n \\ a_1 & a_2 & \ldots & a_n \end{pmatrix}.$

In other words, we apply the permutations from *right to left*.

Note: This is not a standard convention; some authors compose permutations from left to right, so it is necessary to check which convention is being used.

Examples 6.5.2

1. Let $\alpha, \beta \in S_4$ be defined by

$$\alpha = \begin{pmatrix} 1 & 2 & 3 & 4 \\ 4 & 1 & 3 & 2 \end{pmatrix}, \quad \beta = \begin{pmatrix} 1 & 2 & 3 & 4 \\ 2 & 3 & 4 & 1 \end{pmatrix}.$$

We find   (a) $\alpha \circ \beta$;   (b) $\beta \circ \alpha$;   (c) α^2.

It may help you to put in an extra line of working and arrange your calculations (or thoughts) thus:

(a) $\alpha \circ \beta$:

X:	1	2	3	4
$\beta(X)$:	2	3	4	1
$\alpha(\beta(X))$:	1	3	2	4.

Hence

$$\alpha \circ \beta = \begin{pmatrix} 1 & 2 & 3 & 4 \\ 1 & 3 & 2 & 4 \end{pmatrix}.$$

(b) $\beta \circ \alpha$:

X:	1	2	3	4
$\alpha(X)$:	4	1	3	2
$\beta(\alpha(X))$:	1	2	4	3.

Hence

$$\beta \circ \alpha = \begin{pmatrix} 1 & 2 & 3 & 4 \\ 1 & 2 & 4 & 3 \end{pmatrix}.$$

(c) α^2:

X:	1	2	3	4
$\alpha(X)$:	4	1	3	2
$\alpha(\alpha(X))$:	2	4	3	1.

Hence

$$\alpha^2 = \alpha \circ \alpha = \begin{pmatrix} 1 & 2 & 3 & 4 \\ 2 & 4 & 3 & 1 \end{pmatrix}.$$

2. Let $\alpha, \beta, \gamma \in S_5$ be defined by

$$\alpha = \begin{pmatrix} 1 & 2 & 3 & 4 & 5 \\ 2 & 5 & 3 & 1 & 4 \end{pmatrix}, \quad \beta = \begin{pmatrix} 1 & 2 & 3 & 4 & 5 \\ 5 & 4 & 1 & 2 & 3 \end{pmatrix}, \quad \gamma = \begin{pmatrix} 1 & 2 & 3 & 4 & 5 \\ 5 & 3 & 2 & 4 & 1 \end{pmatrix}.$$

Find $\alpha \circ \beta \circ \gamma$:

X:	1	2	3	4	5
$\gamma(X)$:	5	3	2	4	1
$\beta(\gamma(X))$:	3	1	4	2	5
$\alpha(\beta(\gamma(X)))$:	3	2	1	5	4.

Hence

$$\alpha \circ \beta \circ \gamma = \begin{pmatrix} 1 & 2 & 3 & 4 & 5 \\ 3 & 2 & 1 & 5 & 4 \end{pmatrix}.$$

Inverse of a permutation

Let $\alpha \in S_n$ be given by

$$\alpha = \begin{pmatrix} 1 & 2 & \ldots & n \\ a_1 & a_2 & \ldots & a_n \end{pmatrix}.$$

Remembering that the inverse mapping α^{-1} is defined by

$$\alpha^{-1}(a_k) = k \Leftrightarrow \alpha(k) = a_k,$$

we have
$$\alpha^{-1} = \begin{pmatrix} a_1 & a_2 & \ldots & a_n \\ 1 & 2 & \ldots & n \end{pmatrix}.$$

Example 6.5.3

We find the inverse of the permutations α, β of example 6.5.2, 1.

$$\alpha^{-1} = \begin{pmatrix} 4 & 1 & 3 & 2 \\ 1 & 2 & 3 & 4 \end{pmatrix} = \begin{pmatrix} 1 & 2 & 3 & 4 \\ 2 & 4 & 3 & 1 \end{pmatrix};$$

$$\beta^{-1} = \begin{pmatrix} 2 & 3 & 4 & 1 \\ 1 & 2 & 3 & 4 \end{pmatrix} = \begin{pmatrix} 1 & 2 & 3 & 4 \\ 4 & 1 & 2 & 3 \end{pmatrix}.$$

We have now shown how to construct another family of finite groups. The following two results about the members of this family are easy to prove:

RESULT 6.5.2   The order of S_n is $n!$

Proof   Let $\alpha \in S_n$. To prove this result, we compute the number of ways in which we could complete the second line of the expression

$$\alpha = \begin{pmatrix} 1 & 2 & \ldots & n \\ \alpha(1) & \alpha(2) & \ldots & \alpha(n) \end{pmatrix}.$$

Clearly, $\alpha(1)$ can be any of the numbers 1. 2, . . . , n, giving n choices for $\alpha(1)$. But $\alpha(2) \neq \alpha(1)$, since α is injective, and hence once $\alpha(1)$ has been chosen there are only $n-1$ choices for $\alpha(2)$. When $\alpha(1)$ and $\alpha(2)$ have been chosen, there are $n-2$ choices for $\alpha(3)$. Continuing in this way, there are finally just two possibilities for $\alpha(n-1)$ and just one for $\alpha(n)$. Hence the total number of possibilities for α is
$$n(n-1)(n-2) \ldots 2.1 = n! \quad \blacksquare$$

RESULT 6.5.3   When $n \geqslant 3$, S_n is non-abelian.

Proof When $n \geqslant 3$, S_n contains the two permutations

$$\alpha = \begin{pmatrix} 1 & 2 & 3 & \ldots & n \\ 2 & 1 & 3 & \ldots & n \end{pmatrix}, \quad \beta = \begin{pmatrix} 1 & 2 & 3 & 4 & \ldots & n \\ 3 & 2 & 1 & 4 & \ldots & n \end{pmatrix},$$

where, if $n > 3$, $\alpha(x) = x$, $\beta(x) = x$, $4 \leqslant x \leqslant n$.
Then

$$\alpha \circ \beta = \begin{pmatrix} 1 & 2 & 3 & 4 & \ldots & n \\ 3 & 1 & 2 & 4 & \ldots & n \end{pmatrix}. \quad \beta \circ \alpha = \begin{pmatrix} 1 & 2 & 3 & 4 & \ldots & n \\ 2 & 3 & 1 & 4 & \ldots & n \end{pmatrix}.$$

Clearly, $\alpha \circ \beta \neq \beta \circ \alpha$, and hence S_n is non-abelian. ∎

Example 6.5.4

The 3! permutations of S_3 are

$$\iota = \begin{pmatrix} 1 & 2 & 3 \\ 1 & 2 & 3 \end{pmatrix}; \quad \alpha_1 = \begin{pmatrix} 1 & 2 & 3 \\ 2 & 3 & 1 \end{pmatrix}; \quad \alpha_2 = \begin{pmatrix} 1 & 2 & 3 \\ 3 & 1 & 2 \end{pmatrix};$$

$$\beta_1 = \begin{pmatrix} 1 & 2 & 3 \\ 1 & 3 & 2 \end{pmatrix}; \quad \beta_2 = \begin{pmatrix} 1 & 2 & 3 \\ 3 & 2 & 1 \end{pmatrix}; \quad \beta_3 = \begin{pmatrix} 1 & 2 & 3 \\ 2 & 1 & 3 \end{pmatrix}.$$

PROBLEMS 6.5.1a

1 Let $\alpha, \beta \in S_5$ be given by

$$\alpha = \begin{pmatrix} 1 & 2 & 3 & 4 & 5 \\ 4 & 5 & 1 & 2 & 3 \end{pmatrix}, \quad \beta = \begin{pmatrix} 1 & 2 & 3 & 4 & 5 \\ 3 & 2 & 1 & 5 & 4 \end{pmatrix}.$$

Find
(a) $\alpha \circ \beta$;   (b) $\beta \circ \alpha$;   (c) β^{-1};   (d) β^2
(e) $o(\beta)$;   (f) α^{-1};   (g) α^4;   (h) $o(\alpha)$.

2 Construct the Cayley table of S_3, using the notation of example 6.5.4. Find the inverse of each permutation.

3 Find the order of each of the permutations in S_3.

4 Find all the permutations $\alpha \in S_4$ such that

(a) $\alpha(1) = 1$;   (b) $\alpha(1) = 2$.

How many permutations in S_n map 1 onto a given integer a, $1 \leq a \leq n$?

5 Let $\alpha = \begin{pmatrix} 1 & 2 & 3 \\ 2 & 1 & 3 \end{pmatrix}$ and $\beta = \begin{pmatrix} 1 & 2 & 3 \\ 3 & 1 & 2 \end{pmatrix}$. Show that $(\alpha \circ \beta)^2 \neq \alpha^2 \circ \beta^2$.

Cycle notation

There is a more economical notation for expressing permutations which also allows us to see more clearly the structure of the permutation. Let $A = \{x_1, x_2, \ldots, x_k\}$ be a subset of $X = \{1, 2, \ldots, n\}$ and suppose also that $\alpha \in S_n$ is the permutation which maps:

$$x_1 \mapsto x_2, x_2 \mapsto x_3, \ldots, x_{k-1} \mapsto x_k, x_k \mapsto x_1; x \mapsto x, \forall\, x \notin A.$$

We can illustrate the effect of α on the set A as in fig. 6.1.

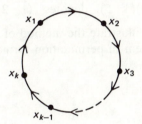

Fig. 6.1

We say that α permutes the numbers in the subset A in a *k-cycle* and that α *fixes* the numbers in X which are not in A. We express this by writing

$$\alpha = (x_1 x_2 \ldots x_k).$$

Examples 6.5.5

1. Let $\alpha \in S_6$ be given by $\alpha = (1 \ \ 4 \ \ 2 \ \ 6)$.
   This tells us that α maps:

   $$1 \mapsto 4, \quad 4 \mapsto 2, \quad 2 \mapsto 6, \quad 6 \mapsto 1 \quad \text{and} \quad 3 \mapsto 3, \ 5 \mapsto 5,$$

   since α fixes any number which does not appear in the cycle. In permutation notation,
   $$\alpha = \begin{pmatrix} 1 & 2 & 3 & 4 & 5 & 6 \\ 4 & 6 & 3 & 2 & 5 & 1 \end{pmatrix}.$$

2. Let $\beta \in S_6$ be given by $(2 \ \ 6 \ \ 1 \ \ 4)$. Does $\alpha = \beta$, where α is the cycle $(1 \ \ 4 \ \ 2 \ \ 6)$ of the previous example?

Notice that since $\alpha, \beta \in S_6$, both permutations have the same domain and codomain. Also β maps:

$$1 \mapsto 4, \quad 4 \mapsto 2, \quad 2 \mapsto 6, \quad 6 \mapsto 1 \quad \text{and} \quad 3 \mapsto 3, \quad 5 \mapsto 5.$$

Hence $\beta = \alpha$.

Thus a cycle can be written starting with any of the elements it permutes.

Composition of cycles

The composition of two or more cycles is sometimes referred to as their *product* and frequently written without the composition symbol. Thus

$$(1\ \ 2\ \ 3)(4\ \ 5) \quad \text{denotes} \quad (1\ \ 2\ \ 3) \circ (4\ \ 5).$$

In examples 6.5.6 we illustrate the method of finding the product of two cycles by rewriting them in permutation form.

Examples 6.5.6

1. (a) $(2\ \ 3\ \ 5)(1\ \ 5\ \ 4) = \begin{matrix} 1 & 2 & 3 & 4 & 5 \\ 5 & 2 & 3 & 1 & 4 \\ 2 & 3 & 5 & 1 & 4 \end{matrix} = \begin{pmatrix} 1 & 2 & 3 & 4 & 5 \\ 2 & 3 & 5 & 1 & 4 \end{pmatrix}.$

   (b) $(1\ \ 5\ \ 4)(2\ \ 3\ \ 5) = \begin{matrix} 1 & 2 & 3 & 4 & 5 \\ 1 & 3 & 5 & 4 & 2 \\ 5 & 3 & 4 & 1 & 2 \end{matrix} = \begin{pmatrix} 1 & 2 & 3 & 4 & 5 \\ 5 & 3 & 4 & 1 & 2 \end{pmatrix}.$

2. (a) $(1\ \ 2\ \ 3)(4\ \ 5) = \begin{matrix} 1 & 2 & 3 & 4 & 5 \\ 1 & 2 & 3 & 5 & 4 \\ 2 & 3 & 1 & 5 & 4 \end{matrix} = \begin{pmatrix} 1 & 2 & 3 & 4 & 5 \\ 2 & 3 & 1 & 5 & 4 \end{pmatrix}.$

   (b) $(4\ \ 5)(1\ \ 2\ \ 3) = \begin{matrix} 1 & 2 & 3 & 4 & 5 \\ 2 & 3 & 1 & 4 & 5 \\ 2 & 3 & 1 & 5 & 4 \end{matrix} = \begin{pmatrix} 1 & 2 & 3 & 4 & 5 \\ 2 & 3 & 1 & 5 & 4 \end{pmatrix}.$

3. $\qquad (1\ \ 2\ \ 3)^2 = \begin{matrix} 1 & 2 & 3 \\ 2 & 3 & 1 \\ 3 & 1 & 2 \end{matrix} = \begin{pmatrix} 1 & 2 & 3 \\ 3 & 1 & 2 \end{pmatrix}.$

Disjoint cycles

Two cycles are said to be *disjoint* if they act on disjoint sets of symbols. Thus in examples 6.5.6 above, the cycles (1 2 3) and (4 5) are disjoint, while the cycles (1 5 4) and (2 3 5) are not. Three or more cycles are said to be *pairwise* disjoint if each pair of the cycles is disjoint. For example, the cycles (1 5)(2 3)(4 6) are pairwise disjoint, whereas the cycles (1 5)(2 3)(4 5) are not, since the symbol 5 is moved by both the first and third cycles.

In examples 6.5.6, 2, we have shown that (1 2 3)(4 5) = (4 5)(1 2 3), and it is not hard to see from the working why they commute. Since the cycles (1 2 3) and (4 5) are disjoint, they act, in a sense, independently of one another and so it does not matter which permutation you consider to be taking place first. Figure 6.2 illustrates the situation.

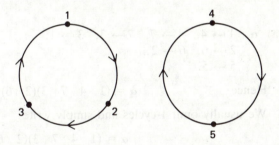

Fig. 6.2

Generalising, we can prove the following result (see problem 6.5.2a, 5).

RESULT 6.5.4.   A pair of disjoint cycles commute.

Notice that cycles which are *not* disjoint do not necessarily (or even usually) commute; example 6.5.6, 1, provides a counter–example.

It is very useful to be able to express a permutation as a product of disjoint cycles, because then its structure is immediately clear. In the following examples we illustrate how this can always be done.

175

Examples 6.5.7

1. We write the permutation $\alpha = \begin{pmatrix} 1 & 2 & 3 & 4 & 5 & 6 \\ 2 & 5 & 6 & 3 & 1 & 4 \end{pmatrix}$ as a product of disjoint cycles.
   Start with any number, say 1. Notice that

   $$\alpha: 1 \mapsto 2, \quad 2 \mapsto 5, \quad 5 \mapsto 1.$$

   Thus (1 2 5) is one of the cycles of which α is composed.
   Next take any of the remaining numbers, say 3. Then

   $$\alpha: 3 \mapsto 6, \quad 6 \mapsto 4, \quad 4 \mapsto 3.$$

   Hence $\alpha = (1 \;\; 2 \;\; 5)(3 \;\; 6 \;\; 4) = (3 \;\; 6 \;\; 4)(1 \;\; 2 \;\; 5)$.

2. We write the permutations

   (a) $\alpha = \begin{pmatrix} 1 & 2 & 3 & 4 & 5 & 6 & 7 \\ 4 & 6 & 1 & 7 & 5 & 2 & 3 \end{pmatrix}$;   (b) $\beta = \begin{pmatrix} 1 & 2 & 3 & 4 & 5 & 6 & 7 \\ 7 & 6 & 1 & 5 & 4 & 2 & 3 \end{pmatrix}$,

   as the product of disjoint cycles.

   (a) $\alpha : 1 \mapsto 4, \quad 4 \mapsto 7, \quad 7 \mapsto 3, \quad 3 \mapsto 1;$
   $ 2 \mapsto 6, \quad 6 \mapsto 2;$
   $ 5 \mapsto 5.$

   Hence $ \alpha = (1 \;\; 4 \;\; 7 \;\; 3)(2 \;\; 6)(5).$

   We usually omit 1-cycles and simply write

   $$\alpha = (1 \;\; 4 \;\; 7 \;\; 3)(2 \;\; 6).$$

   (b) $\beta : 1 \mapsto 7, \quad 7 \mapsto 3, \quad 3 \mapsto 1;$
   $ 2 \mapsto 6, \quad 6 \mapsto 2;$
   $ 4 \mapsto 5, \quad 5 \mapsto 4.$
   Thus $\beta = (1 \;\; 7 \;\; 3)(2 \;\; 6)(4 \;\; 5).$

3. We express the permutation $\alpha = (1 \;\; 3 \;\; 2 \;\; 5)(1 \;\; 2 \;\; 4 \;\; 6)(3 \;\; 6)$ as the product of disjoint cycles.

   Method 1 We first use the method of examples 6.5.6 to write α in permutation form and then the method outlined above to rewrite α as the product of disjoint cycles.
   First, then, remembering to apply the cycles from *right* to *left*,

$$\alpha = \begin{matrix} 1 & 2 & 3 & 4 & 5 & 6 \\ 1 & 2 & 6 & 4 & 5 & 3 \\ 2 & 4 & 1 & 6 & 5 & 3 \\ 5 & 4 & 3 & 6 & 1 & 2 \end{matrix} = \begin{pmatrix} 1 & 2 & 3 & 4 & 5 & 6 \\ 5 & 4 & 3 & 6 & 1 & 2 \end{pmatrix}$$

$$= (1 \quad 5)(2 \quad 4 \quad 6)(3)$$
$$= (1 \quad 5)(2 \quad 4 \quad 6).$$

Method 2 This method is really just a contraction of method 1. To make the explanation clearer, let us denote $(3 \quad 6)$ by β, $(1 \quad 2 \quad 4 \quad 6)$ by γ and $(1 \quad 3 \quad 2 \quad 5)$ by δ. Then $\alpha = \delta\gamma\beta$ and hence $\alpha(x) = \delta(\gamma(\beta(x)))$, $\forall \ x \in X = \{1, 2, 3, 4, 5, 6\}$.
Consider each value of x in turn. Let us start with $x = 1$.

Then, $1 \overset{\beta}{\mapsto} 1 \overset{\gamma}{\mapsto} 2 \overset{\delta}{\mapsto} 5$; thus $\alpha: 1 \mapsto 5$.

Next, $5 \overset{\beta}{\mapsto} 5 \overset{\gamma}{\mapsto} 5 \overset{\delta}{\mapsto} 1$; thus $\alpha: 5 \mapsto 1$.

Hence $(1 \quad 5)$ is one of the disjoint cycles of which α is composed. Now consider any of the remaining numbers, say 2.

$2 \overset{\beta}{\mapsto} 2 \overset{\gamma}{\mapsto} 4 \overset{\delta}{\mapsto} 4$; thus $\alpha: 2 \mapsto 4$;
$4 \overset{\beta}{\mapsto} 4 \overset{\gamma}{\mapsto} 6 \overset{\delta}{\mapsto} 6$; thus $\alpha: 4 \mapsto 6$;
$6 \overset{\beta}{\mapsto} 3 \overset{\gamma}{\mapsto} 3 \overset{\delta}{\mapsto} 2$; thus $\alpha: 6 \mapsto 2$;

and hence $(2 \quad 4 \quad 6)$ is another of the disjoint cycles in α.
Finally, $3 \overset{\beta}{\mapsto} 6 \overset{\gamma}{\mapsto} 1 \overset{\delta}{\mapsto} 3$; thus $\alpha: 3 \mapsto 3$.

Hence $\alpha = (1 \quad 5)(2 \quad 4 \quad 6)(3) = (1 \quad 5)(2 \quad 4 \quad 6)$.

4. We write the permutation $\alpha = (1 \quad 4 \quad 2)(2 \quad 3 \quad 5)(1 \quad 3 \quad 4)$ as a product of disjoint cycles.

Using method 2, we see that

$\alpha:$		
$1 \mapsto 3 \mapsto 5 \mapsto 5,$	giving	$\alpha = (1, 5 \ldots$
$5 \mapsto 5 \mapsto 2 \mapsto 1,$		$\alpha = (1, 5)(\ldots$
$4 \mapsto 1 \mapsto 1 \mapsto 4,$		$\alpha = (1 \quad 5)(4)(\ldots$
$2 \mapsto 2 \mapsto 3 \mapsto 3,$		$\alpha = (1 \quad 5)(4)(2 \quad 3 \ldots$
$3 \mapsto 4 \mapsto 4 \mapsto 2,$		$\alpha = (1 \quad 5)(4)(2 \quad 3).$

Thus $\alpha = (1 \ 5)(2 \ 3)$.
With practice, the steps on the left can be done mentally.

THEOREM 6.5.5 Every permutation is either a cycle or can be written as a product of pairwise disjoint cycles. This expression is unique except for the order in which the cycles appear and for the inclusion or omission of 1-cycles.

Proof This proof generalises the method of example 6.5.7, 1, and uses the generalised form of the principle of induction. The structure of the proof is very similar to that of the unique prime factorisation theorem, theorem 4.5.3. You will find it worthwhile to compare these two proofs.

Let $P(n)$ be the statement that the theorem is true for every permutation on n symbols.

(i) The identity permutation is the only permutation on 1 symbol and this can be written uniquely as the 1−cycle (1). Hence $P(1)$ is trivially true.

(ii) Suppose $P(n)$ is true for all n where $1 \leqslant n < k$. Consider

$$\alpha = \begin{pmatrix} 1 & 2 & 3 & \ldots & k \\ \alpha(1) & \alpha(2) & \alpha(3) & \ldots & \alpha(k) \end{pmatrix}.$$

Then since α is surjective, we can find r, $1 \leqslant r \leqslant k$, such that $\alpha(r) = 1$. Consider the chain

$$1 \overset{\alpha}{\mapsto} a \overset{\alpha}{\mapsto} b \overset{\alpha}{\mapsto} \ldots r \overset{\alpha}{\mapsto} 1.$$

Because α is injective, the numbers $1, a, b, \ldots, r$ in this chain are all distinct. Hence α contains the cycle $\alpha_0 = (1\ a\ b\ \ldots\ r)$. Let $X = \{1, a, b, \ldots, r\}$ be the set of elements permuted by this cycle and $S = \{1, 2, \ldots, n\}$. Then if $X = S$, $\alpha = \alpha_0$ and the theorem is proved. Next suppose $X \subset S$ and let α' be the restriction of α to $S-X$ as domain. Now because α is injective and Im $\alpha_0 = X$, we must have Im $\alpha' = S-X$ and hence α' is a permutation of $S-X$. But $|X| \geqslant 1$ and so $|S-X| < k$. Hence by the induction hypothesis, α' can be expressed uniquely (apart from the order in which the cycles are written) as the product of pairwise disjoint cycles, including all 1-cycles. Let $\alpha' = \alpha_1 \alpha_2 \ldots \alpha_s$ be such an expression for α'. Then $\alpha = \alpha_0 \alpha_1 \alpha_2 \ldots \alpha_s$ is a unique expression for α in the required form, including all 1-cycles. Hence $p(n)$ is true for all $1 \leqslant n < k \Rightarrow P(k)$ is also true. But $P(1)$ is true and hence $P(n)$ is true for all $n \in \mathbb{Z}^+$. ∎

The order of a permutation

When we speak of the *order* of a permutation α, we are regarding α as an element of one of the symmetric groups S_n, for suitable n. We are therefore looking for the least positive integer r such that $\alpha^r = \iota$, where ι denotes the identity permutation.

Example 6.5.8

1. Find the order of the 5-cycle $\alpha = (1 \quad 2 \quad 3 \quad 4 \quad 5)$.
   We are required to find the least integer $r > 0$ such that $\alpha^r(x) = x$, $x = 1, 2, \ldots, 5$. Let us start with $x = 1$.

$$1 \overset{\alpha}{\mapsto} 2 \overset{\alpha}{\mapsto} 3 \overset{\beta}{\mapsto} 4 \overset{\beta}{\mapsto} 5 \overset{\beta}{\mapsto} 1.$$

Hence $\alpha^5(1) = 1$, and it is clear that $\alpha^5(x) = x$, $x = 2, 3, 4, 5$. Also, $\alpha^r(1) \neq 1$ when $0 < r < 5$, and hence $o(\alpha) = 5$.

Generalising example 6.5.8 gives the following result.

RESULT 6.5.6 Let $\alpha = (x_1 x_2 \ldots x_k)$ be a k-cycle. Then α has order k.

RESULT 6.5.7 Let α be the product of two disjoint cycles, $\beta = (x_1 x_2 \ldots x_k)$ and $\gamma = (y_1 y_2 \ldots y_l)$. Then $o(\alpha) = \text{lcm}(k, l)$.

Proof Since β and γ are disjoint, $\beta\gamma = \gamma\beta$, by result 6.5.4.
Hence for any integer r,

$$\alpha^r = (\beta\gamma)^r = \beta\gamma\beta\gamma \ldots \beta\gamma = \beta^r\gamma^r.$$

Let $m = \text{lcm}(k,l)$. Then $k|m$ and $l|m$ and hence $\beta^m = \iota = \gamma^m$, by result 6.3.3. Thus $o(\alpha)|m$.
Also, $\alpha^r = \iota \Rightarrow \beta^r = \iota$ and $\gamma^r = \iota$, since the cycles β and γ are disjoint. But $\beta^r = \iota \Rightarrow k|r$ and $\gamma^r = \iota \Rightarrow l|r$. Thus $k|o(\alpha)$, $l|o(\alpha)$ and hence $m|o(\alpha)$. This, taken with $o(\alpha)|m$, proves

$$o(\alpha) = m. \quad \blacksquare$$

We can extend result 6.5.7 to show that if α is the product of any number of pairwise disjoint cycles, then $o(\alpha)$ is the least common multiple of their cycle lengths. This result, combined with theorem 6.5.5, gives a method for finding the order of any permutation.

Examples 6.5.9

1. We find the order of the permutation $\alpha = \begin{pmatrix} 1 & 2 & 3 & 4 & 5 & 6 \\ 4 & 6 & 1 & 3 & 5 & 2 \end{pmatrix}$.

   Writing α in disjoint cycle form gives

   $$\alpha = (1\ \ 4\ \ 3)(2\ \ 6)(5).$$

   The l.c.m. of the cycle lengths 3, 2, 1 is 6.

   Hence $\qquad\qquad\qquad\qquad o(\alpha) = 6.$

2. We find the order of the permutation $\alpha = (1\ \ 3\ \ 5)(2\ \ 5\ \ 6)(3\ \ 4\ \ 6)$.

   Since the cycles are *not* disjoint, we must first express α as the product of disjoint cycles. This gives:

   $$\alpha = (1\ \ 3\ \ 4\ \ 2)(5\ \ 6).$$

   Hence $\qquad\qquad o(\alpha) = \text{lcm}(4, 2) = 4.$

PROBLEMS 6.5.2a

1 Express each of the following permutations as a single cycle or as the product of disjoint cycles, omitting 1-cycles.

(a) $\begin{pmatrix} 1 & 2 & 3 & 4 & 5 \\ 3 & 5 & 4 & 1 & 2 \end{pmatrix}$;

(b) $\begin{pmatrix} 1 & 2 & 3 & 4 & 5 & 6 & 7 \\ 4 & 6 & 7 & 3 & 2 & 5 & 1 \end{pmatrix}$;

(c) $\begin{pmatrix} 1 & 2 & 3 & 4 & 5 & 6 & 7 \\ 6 & 4 & 5 & 7 & 3 & 1 & 2 \end{pmatrix}$;

(d) $\begin{pmatrix} 1 & 2 & 3 & 4 & 5 & 6 & 7 & 8 \\ 5 & 8 & 7 & 1 & 3 & 6 & 2 & 4 \end{pmatrix}$.

2 Give the order of each of the permutations in problem 1 above.

3 Give the order of each of the following permutations:

(a) $(1\ \ 2\ \ 5\ \ 3\ \ 6)$; (b) $(2\ \ 5)(3\ \ 6)(1\ \ 4)$; (c) $(1\ \ 2\ \ 5\ \ 3)(4\ \ 6)$;

(d) $(1\ \ 2\ \ 3\ \ 4\ \ 5)(6\ \ 7)$; (e) $(1\ \ 2\ \ 5)(3\ \ 4\ \ 6)$.

4 Express each of the following permutations as single cycles or as the product of disjoint cycles. Hence find the order of each.

(a) $(1\ \ 2\ \ 5)(2\ \ 3\ \ 6)$; (b) $(2\ \ 3\ \ 4)(1\ \ 3\ \ 5)(2\ \ 5)$;

(c) $(1\ \ 3\ \ 2\ \ 6)(1\ \ 2\ \ 4)(3\ \ 5)$; (d) $(3\ \ 5\ \ 6)(1\ \ 6)(2\ \ 3\ \ 4)$.

5 Let $\beta = (x_1 x_2 \ldots x_k)$, $\gamma = (y_1 y_2 \ldots y_l)$ be disjoint cycles. By writing each of the products $\beta\gamma$ and $\gamma\beta$ as a permutation on the set $\{x_1, x_2, \ldots, x_k, y_1, y_2, \ldots, y_l\}$, show that $\beta\gamma = \gamma\beta$.

6 Find all the 3-cycles in S_4 that are distinct as permutations.

7 Show that the product (1 6)(1 5)(1 4)(1 3)(1 2) can be expressed as a single cycle.
Suggest a way of expressing the cycle $(x_1 x_2 \ldots x_k)$ as the product of 2-cycles (more than one solution exists).

8 Write all the non-identity elements of S_3 as cycles, omitting 1-cycles.

9 Find the inverse in S_5 of the 5-cycle (1 2 3 4 5). Find also the inverse of the k-cycle $(x_1 x_2 \ldots x_k)$.

10 Give the inverse of

(a) (1 3 2 4); (b) (1 2)(3 4 5); (c) (1 2 3)(3 4 5).

Transpositions

In theorem 6.5.5, we showed that it was possible to express every permutation as a product of disjoint cycles. We have seen that this is useful because it analyses the action of the permutation on its domain and also for calculating the order of the permutation. We now look at another type of decomposition of a permutation which has useful applications, particularly to the theory of determinants and to the solution of polynomial equations. In this decomposition we express every permutation as a product of 2-cycles. A 2-cycle is usually called a *transposition*.

It is easy to see that every cycle can be expressed as a product of transpositions. For example,

$$(x_1 x_2 \ldots x_k) = (x_1 x_k)(x_1 x_{k-1}) \ldots (x_1 x_3)(x_1 x_2)$$

(see problem 6.5.2a, 7). Combining this with theorem 6.5.5 gives the following result:

RESULT 6.5.8 Every permutation is itself a transposition or can be expressed as a product of transpositions.

Examples 6.5.10

1. The following products of transpositions all give (1 2 3 4);

$$(1 \quad 4)(1 \quad 3)(1 \quad 2); \quad (1 \quad 2)(2 \quad 3)(3 \quad 4);$$

$$(1 \quad 4)(3 \quad 4)(2 \quad 4)(3 \quad 4)(1 \quad 3);$$

$$(3 \quad 4)(1 \quad 3)(1 \quad 2)(1 \quad 3)(2 \quad 3)(2 \quad 4)(1 \quad 4).$$

2. The following products of transpositions all give (1 2 3):

$$(1 \quad 3)(1 \quad 2); \quad (1 \quad 2)(2 \quad 3); \quad (1 \quad 3)(2 \quad 3)(1 \quad 2)(1 \quad 3);$$

$$(1 \quad 3)(2 \quad 3)(1 \quad 2)(2 \quad 3)(1 \quad 3)(1 \quad 2).$$

It is clear from examples 6.5.10 that the expression of any cycle as a product of transpositions is *not* unique. However there is one feature of all the different factorisations of a given permutation into transpositions that is constant. Observe that in examples 6.5.10, all the given factorisations of (1 2 3) contain an *even* number of transpositions, while all those of (1 2 3 4) contain an *odd* number. This motivates the following definition:

Even and odd permutations

Any permutation which can be expressed as the product of an *even* number of transpositions (in at least one way) is called an *even* permutation; all other permutations are called *odd*.

We will now prove that if $\alpha \in S_n$ is an *even* permutation then *every* expression of α as a product of transpositions contains an *even* number of transpositions. For the proof we need to introduce a polynomial $P_n(x_1, x_2, \ldots, x_n)$ in n variables defined by

$$P_n = \prod_{i<j}(x_j - x_i).$$

where the product is taken over all factors with $1 \leqslant i < j \leqslant n$. Now we can change the polynomial P_n into another polynomial in n variables by applying a given permutation $\alpha \in S_n$ to the suffixes of the x's. We say that α acts on the polynomial P_n by the rule

$$\alpha: \prod_{i<j}(x_j - x_i) \mapsto \prod_{i<j}(x_{\alpha(j)} - x_{\alpha(i)}).$$

Examples 6.5.11

1. $P_2 = (x_2 - x_1); \ P_3 = (x_2 - x_1)(x_3 - x_1)(x_3 - x_2);$
$P_4 = (x_2 - x_1)(x_3 - x_1)(x_3 - x_2)(x_4 - x_1)(x_4 - x_2)(x_4 - x_3).$

2. Let $\alpha = (1 \quad 2)$ and consider the action of α on P_2, P_3 and P_4.

$\alpha: P_2 \mapsto (x_1 - x_2) = -P_2$.

$\alpha: P_3 \mapsto (x_1 - x_2)(x_3 - x_2)(x_3 - x_1) = -P_3$.

$\alpha: P_4 \mapsto (x_1 - x_2)(x_3 - x_2)(x_3 - x_1)(x_4 - x_2)(x_4 - x_1)(x_4 - x_3)$
$$= -P_4.$$

Notice that in each case the effect of α was to change the factor $(x_2 - x_1)$ into $(x_1 - x_2)$, which *changes the sign* of P_n, $n = 2,3,4$; to *interchange* factors $(x_j - x_1)$ and $(x_j - x_2)$, $j = 3, 4$ which clearly has no net effect on P_n, $n = 3, 4$. The factor $(x_4 - x_3)$ is, of course, unchanged by α. Thus the aggregate effect of α was, in each case, to change P_n into $-P_n$.

3. Let $\alpha = (1 \ 4)$ and consider the action of α on P_4.

$\alpha: P_4 \mapsto (x_2 - x_4)(x_3 - x_4)(x_3 - x_2)(x_1 - x_4)(x_1 - x_2)(x_1 - x_3)$
$$= (-1)^5 P_4 = -P_4.$$

Here, α changes the factor $(x_4 - x_1)$ into $(x_1 - x_4)$ which changes the sign of P_n; the factor $(x_3 - x_2)$ remains unchanged. The remaining four factors fall into two pairs: α changes the pair $(x_j - x_1)(x_4 - x_j)$, $j = 2, 3$ into $(x_j - x_4)(x_1 - x_j) = (x_4 - x_j)(x_j - x_1)$ and thus leaves the product of each pair of factors unchanged. Thus as in (2) above, the aggregate effect of α is to change P_n into $-P_n$.

4. Let $\alpha = (1 \ 2 \ 4)$ and consider the action of α on P_4.

$\alpha: P_4 \mapsto (x_4 - x_2)(x_3 - x_2)(x_3 - x_4)(x_1 - x_2)(x_1 - x_4)(x_1 - x_3)$
$$= (-1)^4 P_4 = P_4.$$

We could have achieved this result another way. Since $\alpha = (1 \ 2 \ 4) = (1 \ 4)(1 \ 2)$, the action of α on P_4 is the same as first applying the transposition $(1 \ 2)$, which changes P_4 into $-P_4$, and then applying $(1 \ 4)$ which also changes P_4 into $-P_4$. Hence the action of α leaves P_4 unchanged.

We generalise the analysis contained in examples 6.5.11, 2 and 3, to prove the following result:

> RESULT 6.5.9   Let $\alpha \in S_n$ be a transposition. Then
> $$\alpha: P_n \mapsto -P_n.$$

Proof   Suppose that $\alpha = (r \ s)$, where $1 \leqslant r \leqslant s \leqslant n$. Clearly the action of α leaves unchanged any factor $(x_j - x_i)$ of P_n in which neither i nor j is r or s. This leaves the following types of factors to consider:

Type (a) Suppose $i < r < s$. Then α *interchanges* the pair of factors $(x_r - x_i)$ and $(x_s - x_i)$.

Type (b) Suppose $r < s < i$. Then α *interchanges* the pair of factors $(x_i - x_r)$ and $(x_i - x_s)$.

Type (c) Suppose $r < i < s$. Then α changes the *pair* of factors $(x_i - x_r)(x_s - x_i)$ into $(x_i - x_s)(x_r - x_i) = (x_s - x_i)(x_i - x_r)$, and thus leaves the *product* of the pair unchanged.

Type (d) The factor $(x_s - x_r)$ is changed by α into $(x_r - x_s) = (-1)(x_s - x_r)$.
Hence the aggregate effect of α is to change P_n into $-P_n$. ∎

From this result, we can immediately deduce the following corollary. It generalises the argument we have used in example 6.5.11, 4.

COROLLARY 6.5.10 Suppose a permutation α can be expressed as the product of r transpositions. Then

$$\alpha: P_n \mapsto (-1)^r P_n.$$

THEOREM 6.5.11 Let α be an even permutation. Then any expression of α as a product of transpositions contains an even number of transpositions.

Proof Since α is even, there is at least one way of expressing α as the product of an even number, say $2r$, of transpositions. Hence by corollary 6.5.10, $\alpha: P_n \to (-1)^{2n} P_n = P_n$. Suppose some other expression of α as the product of transpositions contains s transpositions. Then by corollary 6.5.10, $\alpha: P_n \to (-1)^s P_n$ and hence $(-1)^s = 1$ and s is even. ∎

Note 1: Since $(1 \ \ 2)(1 \ \ 2) = \iota$, the identity is an *even* permutation in S_n, $n \geqslant 2$.

Note 2: It is implicit in the *definition* we gave of an *odd* permutation α that every expression of α as a product of transpositions contains an *odd* number of transpositions.

The result below follows directly from theorem 6.5.11. We refer to it in the next section.

RESULT 6.5.12 (a) The product of any two even or of any two odd permutations is even;
 (b) the product of an odd and an even permutation is odd;
 (c) the inverse of an even permutation is even and of an odd permutation is odd.

Proof See problem 6.5.3a, 4. ■

PROBLEMS 6.5.3a

1 Decide which of the following permutations are even and which are odd.

(a) $(1\ 3\ 2\ 5)$;   (b) $(1\ 2\ 4)(3\ 5)$;
(c) $(1\ 2\ 4)(2\ 3\ 5)$;   (d) $(1\ 3\ 2\ 5)(2\ 4)(3\ 4\ 5)$.

2 (a) Show that a k-cycle is an even permutation when k is odd and an odd permutation when k is even.
 (b) Give an example of a permutation in S_4 which is

   (i) an even permutation with odd order;
   (ii) an even permutation with even order;
   (iii) an odd permutation with even order.

   Is it possible to find an *odd* permutation with *odd* order in S_4? Explain.

3 Express as a transposition or as a product of transpositions, the inverse of each of the following permutations:

(a) $(1\ 2)$;   (b) $(1\ 3)(1\ 2)$;   (c) $(2\ 4)(1\ 3)(1\ 4)(2\ 3)$;

(d) $\beta_1\beta_2 \ldots \beta_k$, where β_i is a *transposition*, $i = 1, 2, \ldots, k$.

4 Prove result 6.5.12.

PROBLEMS 6.5b

1 Show, by giving a concrete example of each, that S_n contains both an odd and an even permutation, $n \geq 2$. Let $\{\alpha_1, \alpha_2, \ldots, \alpha_r\}$ be the odd permutations and $\{\beta_1, \beta_2, \ldots, \beta_s\}$ be the even permutations in S_n, $n \geq 2$. Let α be any odd permutation in S_n. Prove that

(a) $\{\alpha\alpha_1, \alpha\alpha_2, \ldots, \alpha\alpha_r\}$ are distinct even permutations;
(b) $\{\alpha\beta_1, \alpha\beta_2, \ldots, \alpha\beta_s\}$ are distinct odd permutations.

Hence prove that exactly half the permutations in S_n are even and half are odd.

2 Suppose that $\alpha = \alpha_1\alpha_2 \ldots \alpha_r$, where α_i is a cycle of length k_i, $i = 1, 2, \ldots, r$. Prove that α is even or odd depending on whether $k_1 + k_2 + \ldots + k_r - r$ is even or odd. Test each of the permutations in problems 6.5.3a, 1, by this method.

3 Suppose that $\alpha \in S_n$ is a permutation of odd order and let $\alpha = \alpha_1\alpha_2 \ldots \alpha_r$ be an expression for α as a product of pairwise disjoint cycles, where the cycle α_i has length k_i, $i = 1, 2, \ldots, r$. Explain why the integers $k_1, k_2, \ldots, k_r$ are all *odd*. Hence show that whatever the value of $r \geqslant 1$, $k_1 + k_2 + \ldots + k_r - r$ is always *even*. Deduce from the result proved in problem 2 above that every permutation of odd order is an even permutation.

(Look again at problem 6.5.3a, 2(b), in the light of this result.)

6.6 SUBGROUPS

Let H be a non-empty subset of a group $(G, \circ)$. Then if H also satisfies all the group axioms with respect to $\circ$ we call $(H, \circ)$ a *subgroup* of $(G, \circ)$. When no confusion can arise about the operation, we say simply that H *is a subgroup of G.*

Examples 6.6.1

We have already met a number of examples; here are just a few.

1. The set of even integers, $2\mathbb{Z}$, is a group with respect to addition. Hence $(2\mathbb{Z}, +)$ is a subgroup of $(\mathbb{Z}, +)$ and both are subgroups of $(\mathbb{R}, +)$.

2. The set $\{1, -1, i, -i\}$ is a group with respect to multiplication of complex numbers. Hence it is a subgroup of $(\mathbb{C}^*, \times)$. In fact, the set of complex nth roots of unity is a subgroup of $(\mathbb{C}^*, \times)$, $\forall n \in \mathbb{Z}^+$.

3. The set $\left\{ \begin{pmatrix} 1 & 0 \\ 0 & 1 \end{pmatrix}, \begin{pmatrix} 1 & 0 \\ 0 & -1 \end{pmatrix}, \begin{pmatrix} -1 & 0 \\ 0 & 1 \end{pmatrix}, \begin{pmatrix} -1 & 0 \\ 0 & -1 \end{pmatrix} \right\}$ is a group with respect to matrix multiplication. It is therefore a subgroup of the group $GL(2, \mathbb{R})$ of all 2×2 non-singular matrices with real entries with respect to matrix multiplication. This group has many subgroups; the set $\left\{ \begin{pmatrix} x & 0 \\ 0 & y \end{pmatrix} : x, y \in \mathbb{R}^* \right\}$, for example, is an infinite subgroup of $GL(2\ \mathbb{R})$.

4. The sets $A = \{\overline{0}, \overline{6}\}$, $B = \{\overline{0}, \overline{3}, \overline{6}, \overline{9}\}$, $C = \{\overline{0}, \overline{4}, \overline{8}\}$ and
$D = \{\overline{0}, \overline{2}, \overline{4}, \overline{6}, \overline{8}, \overline{9}, \overline{10}\}$ are all subgroups of $(\mathbb{Z}_{12}, \oplus)$. Further, A is
a subgroup of B and D; C is a subgroup of D.

From these examples, we see that it is possible for an infinite group to
have finite and infinite subgroups. In fact every group has two rather
uninteresting subgroups:

RESULT 6.6.1   Let G be a group with identity e. Then
      (a) G is a subgroup of itself;
      (b) $\{e\}$ is a subgroup of G.

Proper subgroups

Let H be a subgroup of G such that $H \neq G$ and $H \neq \{e\}$, then we call
H a *proper* subgroup of G. The subgroup $\{e\}$ is sometimes called the
trivial subgroup of G.

Conditions for a subgroup

THEOREM 6.6.2   Let H be a non-empty subset of a group $(G, \circ)$.
Then H is a *subgroup* if:

(I) H is closed with respect to $\circ$;
(II) when $h \in H$, then its inverse h^{-1} is also a member of H.

Proof Suppose H satisfies the hypotheses (I) and (II). We must show
that $(H, \circ)$ satisfies the four group axioms.

Axiom 1 The closure axiom is satisfied, by hypothesis (I).

Axiom 2 Since $\circ$ is associative in G, it is true that $a \circ (b \circ c) = (a \circ b) \circ c$,
$\forall\, a, b, c \in G$ and hence for every $a, b, c, \in H$. Thus the operation is
necessarily associative in H.

Axiom 3 Since H is non-empty, there is at least one element, say h, in
H. By hypothesis (II), $h \in H \Rightarrow h^{-1} \in H$, and by hypothesis (I), $h \in H$
and $h^{-1} \in H \Rightarrow hh^{-1} \in H$. But $hh^{-1} = e$, the identity element of G.

Since e satisfies $eh = he = h$, $\forall \in H$, then e is an identity element of H and axiom 3 is satisfied.

Axiom 4 By hypothesis (II), $h^{-1} \in H$ and satisfies $hh^{-1} = e$. Since we have shown that e is the identity element of H as well as of G, then h^{-1} is the inverse of h in H and axiom 4 is satisfied. ■

Thus to test whether a subset H of a group $(G, \circ)$ is a subset we must check that

> (i) H is non-empty,
> (ii) $x, y \in H \Rightarrow xy \in H$, $\forall\, x, y \in H$,
> (iii) $x \in H \Rightarrow x^{-1} \in H$, $\forall\, x \in H$.

We do *not* need to check that $\circ$ is associative in H, as this follows from associativity in G; we do *not* need to check specially that the identity is in H since this follows from the other three conditions. (However, it is often obvious that the identity of G is in H, and this may then be the quickest way of showing that H is non-empty.)

Subgroup generated by an element

Let G be a group and suppose $a \in G$. In section 6.4 we introduced the notation $\langle a \rangle$ to denote the set of powers of the element a, that is:

$$\langle a \rangle = \{a^r : r \in \mathbb{Z}\}.$$

For every element $a \in G$, we can form a subset $\langle a \rangle$ and in the next result we use the subgroup test (theorem 6.6.2) to show that $\langle a \rangle$ is a subgroup of G, for all $a \in G$. We call $\langle a \rangle$ the subgroup generated by a.

> RESULT 6.6.3   Let G be a group and $a \in G$. Then $\langle a \rangle$ is a subgroup of G, for all $a \in G$.

Proof (i) Since $a \in \langle a \rangle$, $\langle a \rangle$ is non-empty;

  (ii) Let $x, y \in \langle a \rangle$. Then from the definition of $\langle a \rangle$, both x and y are powers of a, say $x = a^r$, $y = a^s$.
        Then $xy = a^{r+s} \in \langle a \rangle$, since $r + s \in \mathbb{Z}$.
        Hence $x, y \in \langle a \rangle \Rightarrow xy \in \langle a \rangle$.

  (iii) Let $x \in \langle a \rangle$, say $x = a^r$. Then $x^{-1} = (a^r)^{-1} = a^{-r} \in \langle a \rangle$, since $-r \in \mathbb{Z}$.
        Thus $x \in \langle a \rangle \Rightarrow x^{-1} \in \langle a \rangle$.
        Hence $\langle a \rangle$ is a subgroup of G, by theorem 6.6.2. ■

RESULT 6.6.4   Let G be a group and let $a \in G$ be an element of finite order. Then

$$|\langle a \rangle| = o(a).$$

That is, the *order of an element* is equal to the *order of the subgroup generated by that element.*

Proof   Denote the subgroup $\langle a \rangle$ by H. Then the proof is identical to the proof of result 6.4.1, with H everywhere in place of G. Result 6.4.1 is just a special case of result 6.6.4. ∎

Result 6.6.4 explains the connection between our definitions of the order of an *element* and the order of a *group*. Some authors define the order of an element as the order of the subgroup it generates. Result 6.6.4 establishes that this definition is equivalent to the one we gave in section 6.3.

Combinations of subgroups

Suppose H and K are subgroups of a group G. Then since H, K are subsets of the elements in G they can be combined by any of the binary operations on sets defined in Chapter 1. We show in theorem 6.6.5 below that $H \cap K$ is always a subgroup of G; the cases $H \cup K$ and $H - K$ are investigated in problems 6.6.a.

THEOREM 6.6.5   Let H, K be subgroups of a group G. Then $H \cap K$ is again a subgroup of G.

Proof   (See Fig. 6.3.)
   (i)   Let the identity element of G be e.

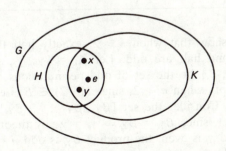

Fig. 6.3

Then since $e \in H$ and $e \in K$, $e \in H \cap K$ and hence $H \cap K$ is non-empty.

(ii)   Let $x, y \in H \cap K$.

Then $x, y \in H$ and hence $xy \in H$, (since H is a subgroup).

Similarly, $x, y \in K$ and hence $xy \in K$. Thus $xy \in H \cap K$.

(iii)   Since $x \in H$, $x^{-1} \in H$, since H is a subgroup.

Similarly, $x \in K \Rightarrow x^{-1} \in K$, and hence $x^{-1} \in H \cap K$.

We have thus shown $H \cap K$ satisfies the three conditions for a subgroup of G. ∎

The alternating groups

In section 6.5, we showed that S_n, $n \leq 2$, can be partitioned into a subset of even and a subset of odd permutations. We denote the subset of *even* permutations in S_n by the symbol A_n where $n \geq 2$. In theorem 6.6.6 we show that A_n is a *subgroup* of S_n, known as the *alternating group* on n symbols.

THEOREM 6.6.6 A_n is a subgroup of S_n, $n \geq 2$.

Proof   (i)   We noted in section 6.5 that the identity permutation ι is an even permutation in S_n, $n \geq 2$, and hence A_n is non-empty.

(ii)   By result 6.5.12(a), the product of any two even permutations is even and hence A_n is closed.

(iii)   By result 6.5.12(c), the inverse of an even permutation is even and hence $\alpha \in A_n \Rightarrow \alpha^{-1} \in A_n$.

Thus A_n is a subgroup of S_n, $n \geq 2$. ∎

THEOREM 6.6.7 $|A_n| = \frac{1}{2}(n!)$, $n \geq 2$.

Proof   We will show that when $n \geq 2$, exactly half the permutations in S_n are even and half are odd. Let $A_n = \{\gamma_1, \gamma_2, \ldots, \gamma_r\}$ and let $B = \{\beta_1, \beta_2, \ldots, \beta_s\}$ be the set of odd permutations in S_n. We know that B is non-empty when $n \geq 2$, since $(1 \ \ 2) \in B$. Let $\beta \in B$ be any odd permutation. Consider the set $\{\beta\gamma_1, \beta\gamma_2, \ldots, \beta\gamma_r\}$. These permutations are distinct, since $\beta\gamma_i = \beta\gamma_j \Rightarrow \gamma_i = \gamma_j$, by theorem 6.2.6. Also, since β is odd and γ_i is even, the product $\beta\gamma_i$ is odd, $i = 1, 2, \ldots, r$, by result 6.5.12(b). Hence the set $\{\beta\gamma_1, \beta\gamma_2, \ldots, \beta\gamma_r\} \subseteq B$ and so $r \leq s$.

Now consider the set $\{\beta\beta_1, \beta\beta_2, \ldots, \beta\beta_s\}$. As above, we can show that these permutations are distinct. Since each is the product of two odd permutations, each permutation in this set is even by result 6.5.12(a).
Hence $\{\beta\beta_1, \beta\beta_2, \ldots, \beta\beta_s\} \subseteq A_n$, implying $s \leq r$.
Thus $r = s$ and $|A_n| = |B| = \frac{1}{2}|S_n|$.
But $|S_n| = n!$, by result 6.5.2, giving the required result. ∎

Examples 6.6.2

The non-identity elements of S_3 can be written in cycle form as

$$\{(1 \quad 2 \quad 3), (1 \quad 3 \quad 2), (1 \quad 2), (2 \quad 3), (1 \quad 3)\}.$$

Now $(1 \quad 2 \quad 3) = (1 \quad 3)(1 \quad 2)$ and $(1 \quad 3 \quad 2) = (1 \quad 2)(1 \quad 3)$ so that these are both *even* permutations.
Thus $A_3 = \{\iota, (1 \quad 2 \quad 3), (1 \quad 3 \quad 2)\}$ and it is easy to verify that A_3 is indeed a subgroup of S_3.

PROBLEMS 6.6a

1 With the notation of example 6.5.4, find the subgroups $\langle \alpha_1 \rangle$ and $\langle \beta_1 \rangle$ of S_3. By writing α_1 and β_1 as cycles, or otherwise, find $o(\alpha_1)$ and $o(\beta_1)$ and verify that $|\langle \alpha_1 \rangle| = o(\alpha_1)$ and $|\langle \beta_1 \rangle| = o(\beta_1)$.
   By constructing a Cayley table, or otherwise, show that the set $\{\iota, \alpha_1, \alpha_2, \beta_1\}$ is not a subgroup of S_3.

2 Suppose that $a \in (G, +)$. Write the definition of the subgroup $\langle a \rangle$ in *additive* notation.
   Find the subgroups $\langle 2 \rangle$ and $\langle 3 \rangle$ of $(\mathbb{Z}, +)$.
   (a) Show that $\langle 2 \rangle \cap \langle 3 \rangle = \langle m \rangle$, giving the value of m.
   (b) Show that $\langle 4 \rangle \cap \langle 6 \rangle = \langle 12 \rangle$. Give a proof of this.
   (c) What is the relationship between a, b and l if $\langle a \rangle \cap \langle b \rangle = \langle l \rangle$?

3 Let $\langle 2 \rangle$, $\langle 3 \rangle$ be the subgroups of $(\mathbb{Z}, +)$ of problem 2.
   Show that $\langle 2 \rangle \cup \langle 3 \rangle$ is *not* a subgroup of $(\mathbb{Z}, +)$.
   Suppose H, K are subgroups of a group G. Decide which of the following statements about $H \cup K$ is true and which are false, briefly substantiating your answer.
   (a) $H \cup K$ is always a subgroup of G;
   (b) $H \cup K$ is never a subgroup of G;
   (c) $H \cup K$ is not necessarily a subgroup of G.

4 Let H, K be subgroups of a group G. Give a reason why $H - K$ is never a subgroup of G.

5 Find the elements of $\langle (1 \ \ 2 \ \ 3 \ \ 4) \rangle$ in the group S_4.

6 Find the subgroup generated by each element of $(\mathbb{Z}_5, \oplus)$. Deduce that there exist groups which have no proper subgroups.

7 Prove that the set $\{\iota, (1 \ \ 2), (3 \ \ 4), (1 \ \ 2)(3 \ \ 4)\}$ is a subgroup of S_4.

8 Find the following subsets of S_3:

(a) $H = \{\alpha^2 : \alpha \in S_3\}$;
(b) $K = \{\alpha^3 : \alpha \in S_3\}$.

Show that H is a subgroup while K is not.

9 Decide, giving reasons, whether each of the following subsets of S_n ($n \geq 2$), is or is not a subgroup:

(a) $\{\alpha \in S_n : \alpha(1) = 1\}$;
(b) $\{\alpha \in S_n : \alpha(1) = 2\}$.

10 Let g be a fixed element of a group G. Let C_g be the subset of elements of G that commute with g. That is, $C_g = \{x \in G : xg = gx\}$. Prove that C_g is a subgroup of G.

11 Let G be an abelian group. Let H be the subset of G consisting of those elements $x \in G$ such that $x = x^{-1}$. That is

$$H = \{x \in G : x = x^{-1}\}.$$

Prove that H is a subgroup of G.

12 List the elements of A_4.

PROBLEMS 6.6b

1 Find the subgroups $\langle \overline{3} \rangle$, $\langle \overline{5} \rangle$, $\langle \overline{6} \rangle$, $\langle \overline{10} \rangle$ in $(\mathbb{Z}_{30}, \oplus)$.
What is the relation between m and n if $\langle \overline{m} \rangle \subseteq \langle \overline{n} \rangle$?

2 By constructing a Cayley table, or otherwise, prove that the set

$$H = \{\iota, (1 \ \ 2)(3 \ \ 4), (1 \ \ 3)(2 \ \ 4), (1 \ \ 4)(2 \ \ 3)\}$$

is a subgroup of S_4.
Find all the subgroups of H. Explain why it is not possible for H to contain a subgroup of order 3.

3 Let $\mathcal{F}$ be a family of subgroups of a group G. Generalise the proof of result 6.6.5 to show that the intersection of all the subgroups in $\mathcal{F}$ is a subgroup of G.

4 Show that the set $\{\iota, (1\ 3\ 2\ 4), (1\ 2)(3\ 4), (1\ 4\ 2\ 3), (1\ 3)(2\ 4)\}$ is not a subgroup of S_4.

5 Decide, giving reasons, whether each of the following subsets of S_n $(n \geq 3)$ is or is not a subgroup of S_n.

(a) $\{\alpha \in S_n: \alpha(3) = 3\}$;
(b) $\{\alpha \in S_n: \alpha(1) = 1 \text{ or } 2\}$;
(c) $\{\alpha \in S_n: \alpha(1) \text{ and } \alpha(2) \text{ are both in } \{1, 2\}\}$.

6 Let G be a group containing subgroups H and K such that we can find an element $h \in H - K$ and an element $k \in K - H$.
Prove that $hk \notin H \cup K$.
Deduce that $H \cup K$ is not a subgroup of G.
Hence prove that $H \cup K$ is a subgroup if and only if $H \subseteq K$ or $K \subseteq H$.

7 Let G be an abelian group. Let r be a fixed positive integer and let K be the subset of G defined by

$$K = \{x^r: x \in G\}.$$

That is, K consists of the rth power of every element of G.
Prove that K is a subgroup of G. Would this result necessarily be true if G were not abelian?

8 Let G be a group and C be the subset of G consisting of just those elements that commute with *every* element of the group. Prove that C is a subgroup of G.

9 Let G be a group and let H be a subset of G satisfying the following conditions:

(i) H is non-empty;
(ii) $\forall a, b \in H, ab^{-1} \in H$.

Prove that H is a subgroup of G.

10 Let $A_n = \{\gamma_1, \gamma_2, \ldots, \gamma_r\}$, where $r = \frac{1}{2}(n!)$, and let α be *any* permutation in S_n. Prove that the set $\{\alpha^{-1}\gamma_i\alpha: 1 \leq i \leq r\} = A_n$.

6.7 GROUPS OF SYMMETRIES

Some of the most important applications of group theory arise from the connection between groups and symmetry. We consider here, as an introduction to the idea, the groups of symmetries of some simple plane geometrical figures.

Symmetry operations

Suppose we have drawn an equilateral triangle ABC and placed an identical triangle cut from card on top of it. Any operation, or combination of operations, on the cardboard triangle of the following types:

(a) a 'flip', or rotation through π radians about an axis in the plane of ABC,

(b) a rotation of the triangle in its own plane about an axis normal to the plane of ABC,

which leaves the cardboard triangle covering ABC, is called a *symmetry operation*, or, more simply, a *symmetry* of the triangle.

There are three symmetries of the first type, as shown in Fig. 6.4; let σ_1 denote a flip about the axis through A bisecting BC; in other words, σ_1 keeps the same vertex on A and interchanges the vertices on B and C.

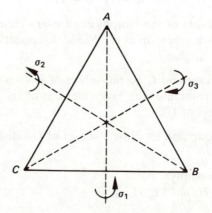

Fig. 6.4

Let σ_2 and σ_3 denote similar flips about the axes through B and C.
There are also rotational symmetries:
Let ρ_1 denote a clockwise rotation of the triangle through an angle of $\dfrac{2\pi}{3}$ about its circumcentre, so that the vertex that was on A moves onto B etc., and let ρ_2 denote a similar clockwise rotation through $\dfrac{4\pi}{3}$, so that

the vertex that was on A moves onto C. A similar rotation through 2π would bring each vertex back to its original position. We denote any symmetry, or combination of symmetries, which leaves the triangle in its original position by ι.

To illustrate the result of performing each of these symmetry operations on the cardboard triangle, we shall number its vertices so that initial vertex 1 is on A, 2 is on B and 3 is on C. Figure 6.5 illustrates the position the triangle is in after performing each of the symmetries ι, σ_1, σ_2, σ_3, ρ_1, ρ_2, defined above. It is easy to check that this list accounts for all possible symmetries of the triangle.

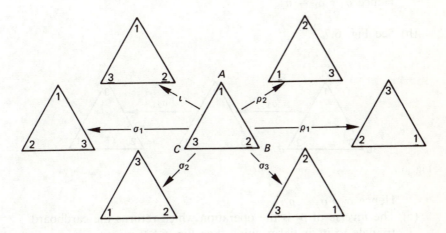

Fig. 6.5

Combinations of symmetries

We combine any two symmetries by performing them one after the other. Since a symmetry is equivalent to a permutation of the set of vertices, we adopt the same notation as for mappings. That is, we write

$$\tau \circ \sigma$$

to mean: *first* perform operation σ and *then* perform τ.

Examples 6.7.1

1. With the notation above for the symmetries of the equilateral triangle, find (a) $\sigma \circ \rho_1$; (b) $\sigma_2 \circ \sigma_3$; (c) ρ_2^{-1}.

Solution

(a) See Fig. 6.6.

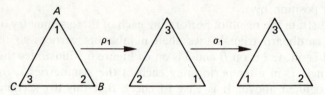

Fig. 6.6

Hence $\sigma_1 \circ \rho_1 = \sigma_2$.

(b) See Fig. 6.7.

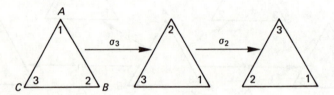

Fig. 6.7

Hence $\sigma_2 \circ \sigma_3 = \rho_1$.

(c) The *inverse* of ρ_2 is the operation which returns the cardboard triangle to its initial position. See Fig. 6.8.

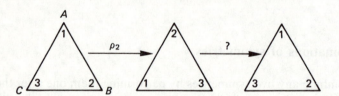

Fig. 6.8

Hence $\rho_2^{-1} = \rho_1$.

2. Write down the permutation on the set of vertices $\{1, 2, 3\}$ which is equivalent to (a) σ_1; (b) ρ_1.

Solution

(a) σ_1 keeps the vertex 1 on A, but interchanges the positions of the vertices 2 and 3. Thus σ_1 is equivalent to the permutation

196

$$\beta_1 = \begin{pmatrix} 1 & 2 & 3 \\ 1 & 3 & 2 \end{pmatrix}.$$

(b) Similarly, ρ is equivalent to $\alpha_1 = \begin{pmatrix} 1 & 2 & 3 \\ 2 & 3 & 1 \end{pmatrix}$, since ρ_1 has moved 1 onto the initial position of 2, 2 onto 3 and 3 onto 1.

The definition given above of a *symmetry operation* on an equilateral triangle is just a special case of the more general idea of a symmetry operation on *any* plane geometrical figure. We have seen that the symmetries of the equilateral triangle are each a rotation about an axis (either in or normal to the plane of the triangle) passing through the circumcentre.

Given any plane geometrical figure F, any rotation about an axis in space that transforms F into itself is called a *symmetry* of F. If F has more than one axis of rotational symmetry, then these axes will intersect in a point O in its plane. The symmetries we have described are called symmetries of F with respect to O.

THEOREM 6.7.1 Let S be the set of all symmetries of a given plane geometrical figure F with respect to an origin O in its plane. Then S is a group with respect to combination of symmetries.

Proof

Axiom 1 The combination of two symmetries of F results in a symmetry of F which must be a member of S, since S includes all the symmetries of F with respect to O.

Axiom 2 That this rule of combination is associative, follows as for composition of mappings.

Axiom 3 The symmetry we have denoted by ι is an identity in S.

Axiom 4 The inverse of a symmetry is the symmetry which returns the figure to its initial position. It is clear that this is in each case a symmetry of the same type i.e. the inverse of a flip is a flip, and the inverse of a rotation is another rotation. By the definition of S, these symmetries will also be members of S.

Symmetry group

The group whose elements are the complete set of symmetries of a given geometrical figure with respect to an origin O in its plane is called the *symmetry group* of that figure.

Dihedral group

The family of symmetry groups of the *regular* polygons are known as the *dihedral groups*. The symmetry group of the triangle is denoted by D_3, of the square by D_4 and in general, the symmetry group of a regular *n*-gon by D_n.

In the applications of group theory to molecular structure and particle physics, the symmetries of 3-dimensional figures are of central importance. Given a polyhedron (such as a cube, for example), any rotation about an axis in space which transforms the polyhedron into itself is called a *symmetry* of the polyhedron. The proof of theorem 6.7.1 is easily adapted to show that the set of rotational symmetries of any 3-dimensional figure form a group with respect to combination of symmetries (see problems 6.7b, 5).

PROBLEMS 6.7a

1 Construct the Cayley table for the group $S = \{\iota, \rho_1, \rho_2, \sigma_1, \sigma_2, \sigma_3\}$ of symmetries of an equilateral triangle, defined above. Show that this group is not abelian.

2 Show that the subset of rotations $\{\iota, \rho_1, \rho_2\}$ is a *subgroup* of S, but that the subset of flips $\{\iota, \sigma_1, \sigma_2, \sigma_3\}$ is not.

3 Write down the permutations $\beta_2, \beta_3, \alpha_2$ of the set of vertices $\{1, 2, 3\}$ which are equivalent to the symmetries σ_2, σ_3 and ρ_2 respectively. Using the rule for multiplying permutations, check that

$$\text{(a) } \beta_1 \circ \alpha_1 = \beta_2; \quad \text{(b) } \beta_2 \circ \beta_3 = \alpha_1; \quad \text{(c) } \alpha_2^{-1} = \alpha_1$$

where α_1 and β_1 are defined in example 6.7.1, 2 above.

4 Check that the symmetries $\iota, \sigma_1, \sigma_2, \rho_2$ defined below are all the possible symmetries of a rectangle with two unequal sides (Fig. 6.9).

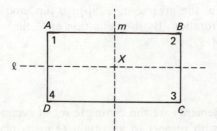

Fig. 6.9

ι identity symmetry;
σ_1 flip about the axis l;
σ_2 flip about the axis m;
ρ_2 rotation through π about the point X.

(a) Draw diagrams to show the result of performing each symmetry.
(b) Construct the Cayley table of the group of symmetries of the rectangle. Show that this group is abelian.
(c) Write down the permutation on the set $\{1, 2, 3, 4\}$ which corresponds to each symmetry. Show that this set of permutations forms a subgroup of S_4.

5 Show that there are just eight symmetries of a square and define them as we have done in problem 4, above, for the rectangle. Draw a diagram to show the result of performing each symmetry.

Construct the Cayley table for the symmetry group D_4 of the square (this group is also known as the *octic group*). Show that it is nonabelian.

6 Find three subgroups of D_4 which have order 4.

7 Give the permutation on the set $\{1, 2, 3, 4\}$ which corresponds to each of the symmetries of a square. Show that they form a subgroup of S_4.

8 With the notation of problem 4, above, show that the group G of symmetries of a rectangle has five subgroups, including the improper ones. Regarding the letters of the alphabet, written in capitals, as geometrical objects, show that they can be classed into five subsets so that the letters of each subset have the same subgroup of G as symmetry group.

PROBLEMS 6.7b

1 Define the symmetries of (a) a rhombus; (b) a parallelogram with two unequal sides. Compare the results with those of problems 6.7a, 4 and 5, for a rectangle and a square. Notice that the more geometrical symmetry a figure possesses, the greater the order of its symmetry group.

2 Show, with the aid of a diagram, that there are 10 symmetries of a regular pentagon.

Show that the subset of rotational symmetries, including the identity, forms a cyclic subgroup of order 5. Denote this subgroup by $\{\iota, \rho_1, \rho_2, \rho_3, \rho_4\}$ and let σ_1 denote any one of the flips. Show that the other four flips are given by $\sigma_1 \circ \rho_1$, $\sigma_1 \circ \rho_2$, $\sigma_1 \circ \rho_3$ and $\sigma_1 \circ \rho_4$.

3 Show that the order of the symmetry group D_6 of a regular hexagon is 12. Find, also, the order of the symmetry group of a hexagon of

Fig. 6.10

which four sides are equal, and one pair of opposite sides are equal to one another but longer than the other four (Fig. 6.10).

4 Show that the order of the symmetry group D_n of a regular plane polygon with n vertices is $2n$, considering the cases where n is even and n is odd separately.

5 (a) Adapt the proof of theorem 6.7.1 to show that the set of symmetries of a 3-dimensional figure F form a group with respect to combination of symmetries.

  (b) Show that a regular tetrahedron has just 4 axes of symmetry. Find the number of symmetry operations which can be performed about each axis. Let S_i be the set of symmetries about axis l_i, $i = 1, 2, 3, 4$, and G be the group of symmetries of the tetrahedron. Show that

    (i) S_i is a subgroup of G, $i = 1, 2, 3, 4$.
    (ii) $S_i \cap S_j = \{\iota\}$, when $1 \leqslant i < j \leqslant 4$.
    Hence find the order of G.

SOLUTIONS TO PROBLEMS IN THE (a) SETS

Problems 6.1.1a

1   (a) Group;   (b) group;   (c) group;
     (d) axioms 3, 4 not satisfied;   (e) axiom 4 not satisfied;
     (f) group;   (g) group;   (h) group;
     (i) and (j) axiom 4 not satisfied.

2 *Closure:* $a, b, \in A \Rightarrow (a, b \in \mathbb{Q}, a, b \neq 1) \Rightarrow a + b - ab \in \mathbb{Q}$.
   Suppose $a + b - ab = 1$. Then $(a - 1)(b - 1) = 0 \Rightarrow (a = 1$ or $b = 1)$. Hence $a, b \in A \Rightarrow a \circ b \in A$.
   Associativity: Show $(a \circ b) \circ c = a + b + c - ab - ac - bc + abc$ $= a \circ (b \circ c)$.
   Identity: Solve $a \circ e = a$ to show that $e = 0$ is an identity.

Inverses: Solve $a \circ a' = e = 0$ to show that every $a \in A$ has an inverse $a' = a/(a - 1) \in A$.

3 *Closure:* $a, b \in \mathbb{Z} \Rightarrow a + b + 1 \in \mathbb{Z}$.
Associativity: Show $(a \circ b) \circ c = a + b + c + 2 = a \circ (b \circ c)$.
Identity: Solve $a \circ e = a$ to show that $e = -1$ is an identity.
Inverses: Solve $a \circ a' = -1$ to show that every $a \in \mathbb{Z}$ has an inverse $a' = -(a + 2) \in \mathbb{Z}$.

4 (a) *Closure:* Let $X = \begin{pmatrix} x & 0 \\ 0 & 0 \end{pmatrix}$, $Y = \begin{pmatrix} y & 0 \\ 0 & 0 \end{pmatrix} \in W_1$. Then

$$XY = \begin{pmatrix} xy & 0 \\ 0 & 0 \end{pmatrix} \in W_1.$$

Associativity: Matrix multiplication is associative.

Identity: $\begin{pmatrix} 1 & 0 \\ 0 & 0 \end{pmatrix}$

Inverse of X is $X^{-1} = \begin{pmatrix} 1/x & 0 \\ 0 & 0 \end{pmatrix} \in W_1, \forall X \in W_1$.

(b) *Closure:* Let $X = \begin{pmatrix} x & -x \\ 0 & 0 \end{pmatrix}$, $Y = \begin{pmatrix} y & -y \\ 0 & 0 \end{pmatrix} \in W_2$. Then $XY = \begin{pmatrix} xy & -xy \\ 0 & 0 \end{pmatrix} \in W_2$.

Associativity: Matrix multiplication is associative.

Identity: $\begin{pmatrix} 1 & -1 \\ 0 & 0 \end{pmatrix}$

Inverse of $X \in W_1$ is $X = \begin{pmatrix} 1/x & -1/x \\ 0 & 0 \end{pmatrix} \in W_1, \forall X \in W_1$.

5 *Closure:* Let $(x_1, y_1), (x_2, y_2) \in S$. Then $x_1, x_2 \in \mathbb{R} \Rightarrow x_1 + x_2 \in \mathbb{R}$ and $y_1, y_2 \in \mathbb{R}^* \Rightarrow y_1 y_2 \in \mathbb{R}^* \Rightarrow (x_1, y_1) \circ (x_2, y_2) \in S$.
Associativity: Show that

$$\{(x_1, y_1) \circ (x_2, y_2)\} \circ (x_3, y_3) = (x_1 + x_2 + x_3, y_1 y_2 y_3)$$
$$= (x_1, y_1) \circ \{(x_2, y_2) \circ (x_3, y_3)\}.$$

Identity: Show that $(0, 1)$ is an identity.
Inverse of $(x_1, y_1) \in S$ is $(-x_1, 1/y_1) \in S$.

Problems 6.1.2a

1 Let $I = \begin{pmatrix} 1 & 0 \\ 0 & 0 \end{pmatrix}$, $A = \begin{pmatrix} -1 & 0 \\ 0 & 1 \end{pmatrix}$, $B = \begin{pmatrix} 1 & 0 \\ 0 & -1 \end{pmatrix}$, $C = \begin{pmatrix} -1 & 0 \\ 0 & -1 \end{pmatrix}$.

×	I	A	B	C
I	I	A	B	C
A	A	I	C	B
B	B	C	I	A
C	C	B	A	I

Closure: From table.
Associativity: Matrix multiplication is associative.
Identity element: I.
Inverses: Each element is its own inverse.
Hence all four axioms are satisfied.

2 $\mathbb{Z}_3$:

$\oplus$	$\bar{0}$	$\bar{1}$	$\bar{2}$
$\bar{0}$	$\bar{0}$	$\bar{1}$	$\bar{2}$
$\bar{1}$	$\bar{1}$	$\bar{2}$	$\bar{0}$
$\bar{2}$	$\bar{2}$	$\bar{0}$	$\bar{1}$

$\mathbb{Z}_4$:

$\oplus$	$\bar{0}$	$\bar{1}$	$\bar{2}$	$\bar{3}$
$\bar{0}$	$\bar{0}$	$\bar{1}$	$\bar{2}$	$\bar{3}$
$\bar{1}$	$\bar{1}$	$\bar{2}$	$\bar{3}$	$\bar{0}$
$\bar{2}$	$\bar{2}$	$\bar{3}$	$\bar{0}$	$\bar{1}$
$\bar{3}$	$\bar{3}$	$\bar{0}$	$\bar{1}$	$\bar{2}$

(a) $\bar{2}$. (b) $\bar{3}$.

3 When $n \geqslant 2$, $\bar{1} \in \mathbb{Z}_n$ and satisfies $\bar{1} \otimes \bar{a} = \bar{a} \otimes \bar{1} = \bar{a}$, $\forall\, \bar{a} \in \mathbb{Z}_n$.

4 $\mathbb{Z}_5$:

$\otimes$	$\bar{0}$	$\bar{1}$	$\bar{2}$	$\bar{3}$	$\bar{4}$
$\bar{0}$	$\bar{0}$	$\bar{0}$	$\bar{0}$	$\bar{0}$	$\bar{0}$
$\bar{1}$	$\bar{0}$	$\bar{1}$	$\bar{2}$	$\bar{3}$	$\bar{4}$
$\bar{2}$	$\bar{0}$	$\bar{2}$	$\bar{4}$	$\bar{1}$	$\bar{3}$
$\bar{3}$	$\bar{0}$	$\bar{3}$	$\bar{1}$	$\bar{4}$	$\bar{2}$
$\bar{4}$	$\bar{0}$	$\bar{4}$	$\bar{3}$	$\bar{2}$	$\bar{1}$

$(\mathbb{Z}_5, \otimes)$ has an identity element $\bar{1}$. But there is no element $\bar{m} \in \mathbb{Z}_5$ such that $\bar{0} \otimes \bar{m} = \bar{1}$. Hence $\bar{0}$ has no inverse and axiom 4 is not satisfied.

Generalising: when $n \geqslant 2$, $\bar{1}$ is the identity element of $(\mathbb{Z}_n, \otimes)$. But $\bar{0} \otimes \bar{a} = \bar{0}$, $\forall\, \bar{a} \in \mathbb{Z}_n$ and hence $\bar{0}$ has no inverse w.r.t. $\otimes$ in $\mathbb{Z}_n$.

5 (a) Group.
  (b) Not a group; axioms 1 and 4 not satisfied.
  (c) Group.
  (d) Not a group; axioms 1 and 4 not satisfied.
  (e) Group.

6 Let $n = ab$, where $1 < a < n$, $1 < b < n$. Then in $\mathbb{Z}_n$,

$$\bar{a} \otimes \bar{b} = \overline{ab} = \bar{n} = \bar{0} \notin \mathbb{Z}_n^*.$$

Thus $(\mathbb{Z}_n^*, \otimes)$ is *not* closed.

7

$\otimes$	$\bar{1}$	$\bar{3}$	$\bar{5}$	$\bar{7}$
$\bar{1}$	$\bar{1}$	$\bar{3}$	$\bar{5}$	$\bar{7}$
$\bar{3}$	$\bar{3}$	$\bar{1}$	$\bar{7}$	$\bar{5}$
$\bar{5}$	$\bar{5}$	$\bar{7}$	$\bar{1}$	$\bar{3}$
$\bar{7}$	$\bar{7}$	$\bar{5}$	$\bar{3}$	$\bar{1}$

Closure: From table.
Associativity: See result 6.1.4.
Identity element: $\bar{1}$.
Inverses: Each element is its own inverse.
Hence all four axioms are satisfied.

8 Show that axiom 1 (or axiom 4) is not satisifed.

9 (a) Let $z = z_1 z_2$. Then $z^n = (z_1 z_2)^n = z_1{}^n z_2{}^n = 1$, since $z_1{}^n = 1$ and $z_2{}^n = 1$.

(b) Let $z = 1/z_1$. Then $z^n = 1/z_1{}^n = 1$, since $z_1{}^n = 1$.
Let $G = \{z \in \mathbb{C}: z^n = 1\}$. Consider $(G, \times)$.
Closure: Follows from (a).
Associativity: Multiplication of complex numbers is associative.
Identity: The complex number $1 + 0i \in G$.
Inverses: Let $z \in G$. Then $1/z \in G$ from (b).
Hence all four group axioms are satisfied by $(G, \times)$.
The group $(H, \times)$ of example 6.1.2 is the group of complex fourth roots of unity; that is $H = \{z \in \mathbb{C}: z^4 = 1\}$.

10 (a) $(f_2 \circ f_3)(x) = f_2(f_3(x)) = f_2(1 - x) = 1/(1 - x) = f_4(x)$
$\Rightarrow f_2 \circ f_3 = f_4$;
(b) $(f_4 \circ f_6)(x) = f_4(f_6(x)) = f_4(1 - 1/x) = 1/(1/x) = f_1(x)$
$\Rightarrow f_4 \circ f_6 = f_1$.

$\circ$	f_1	f_2	f_3	f_4	f_5	f_6
f_1	f_1	f_2	f_3	f_4	f_5	f_6
f_2	f_2	f_1	f_4	f_3	f_6	f_5
f_3	f_3	f_6	f_1	f_5	f_4	f_2
f_4	f_4	f_5	f_2	f_6	f_3	f_1
f_5	f_5	f_4	f_6	f_2	f_1	f_3
f_6	f_6	f_3	f_5	f_1	f_2	f_4

Closure: From table.

Associativity: Composition of mappings is associative.

Identity element: f_1.

Inverses: f_1, f_2, f_3, f_5 is each its own inverse. f_4 and f_6 are inverses of each other.

Hence $(M, \circ)$ satisfies all four group axioms.

Since e.g. $f_2 \circ f_3 = f_4$ and $f_3 \circ f_2 = f_6$, $(M, \circ)$ is *non-abelian*.

11 A group $(G, \circ)$ is abelian $\Leftrightarrow a \circ b = b \circ a$, $\forall a, b \in G$. Thus the Cayley table of an abelian group must be *symmetrical* about the main diagonal line (top left to bottom right). Compare the tables given above for problems 1, 2, 4 and 7 with that for problem 10.

Problems 6.2a

1 (a) Cancellation laws do not hold, since $b \circ a = b \circ c$ (and $a \circ b = c \circ b$). (Or, harder, show that $\circ$ is not associative: e.g. $(b \circ b) \circ c \neq b \circ (b \circ c)$.)

   (b) No identity element.

2

$\circ$	x	y
x	x	y
y	y	x

Since x is the identity, $x \circ x = x$, $x \circ y = y \circ x = y$. By closure axiom, $y \circ y = x$ or y. Using result 6.2.8 (or 6.2.7) gives $y \circ y = x$ as the only possibility.

3 (a) If $AB = AC$, with $B \neq C$, then A must be *singular*. For, if A^{-1} exists, then $AB = AC \Rightarrow A^{-1}(AB) = A^{-1}(AC) \Rightarrow (A^{-1}A)B = (A^{-1}A)C \Rightarrow B = C$, giving a contradiction.

   Choose A as simply as possible: say $A = \begin{pmatrix} 1 & 0 \\ 0 & 0 \end{pmatrix}$. Choose B and C with the same first row, but different second rows, e.g. let $B = \begin{pmatrix} 1 & 2 \\ 3 & 4 \end{pmatrix}$, $C = \begin{pmatrix} 1 & 2 \\ 5 & 6 \end{pmatrix}$. Then $AB = AC = \begin{pmatrix} 1 & 2 \\ 0 & 0 \end{pmatrix}$.

   $M(2, \mathbb{R})$ is *not* a group; axiom 4 does not hold.

   (b) This time A^{-1} exists. Notice that $AB = CA \Rightarrow A^{-1}(AB) = A^{-1}(CA) \Rightarrow (A^{-1}A)B = A^{-1}CA \Rightarrow B = A^{-1}CA$. Now if $CA = AC$, then $B = A^{-1}AC = C$. Hence we must choose C so that $AC \neq CA$. Choose A as simply as possible: say $A = \begin{pmatrix} 1 & 0 \\ 0 & 2 \end{pmatrix}$.

   Then $A^{-1} = \begin{pmatrix} 1 & 0 \\ 0 & \frac{1}{2} \end{pmatrix}$.

As an example, let $C = \begin{pmatrix} 1 & 1 \\ 1 & 0 \end{pmatrix}$. Verify $AC \neq CA$. Then $B =$
$A^{-1}CA = \begin{pmatrix} 1 & 2 \\ \frac{1}{2} & 0 \end{pmatrix}$. Verify $AB = CA = \begin{pmatrix} 1 & 2 \\ 1 & 0 \end{pmatrix}$.

The cancellation law is not violated since A *pre*-multiplies B and *post*-multiplies C.

4 *Injective:* Let $x, y \in G$. Then $\alpha(x) = \alpha(y) \Rightarrow a \circ x = a \circ y \Rightarrow x = y$, by cancellation law.
Surjective: Let g be any element of G. Then α is surjective if we can find $x \in G$ such that $\alpha(x) = g$. But $\alpha(x) = a \circ x$ and $a \circ x = g$ has the (unique) solution $x = a^{-1} \circ g$, by theorem 6.2.7.
Thus α is injective and surjective and hence bijective.

5 First suppose G is abelian. Then $a \circ b = b \circ a$, $\forall a, b \in G$.
Hence $(a \circ b)^2 = (a \circ b) \circ (a \circ b)$
$\qquad\qquad\qquad = a \circ (b \circ a) \circ b$, since $\circ$ is associative,
$\qquad\qquad\qquad = a \circ (a \circ b) \circ b$, since $\circ$ is commutative,
$\qquad\qquad\qquad = (a \circ a) \circ (b \circ b)$, since $\circ$ is associative.

Conversely, if $(a \circ b)^2 = a^2 \circ b^2$, $\forall a, b \in G$, then
$\qquad\qquad (a \circ b) \circ (a \circ b) = (a \circ a) \circ (b \circ b)$,
$\qquad \Rightarrow a \circ (b \circ a) \circ b = a \circ (a \circ b) \circ b$, since $\circ$ is associative,
$\qquad \Rightarrow b \circ a = a \circ b$, $\forall a, b \in G$, by the cancellation laws.

6 $a \circ b \circ c = e \Rightarrow a \circ (b \circ c) = e \Rightarrow b \circ c = a^{-1}$.
But $a^{-1} \circ a = e$ and hence $b \circ c \circ a = e$.
The other way of inserting brackets in $a \circ b \circ c$ gives $(a \circ b) \circ c = e$
$\Rightarrow a \circ b = c^{-1} \Rightarrow c \circ (a \circ b) = e \Rightarrow c \circ a \circ b = e$.
The other combinations do not necessarily give the identity.

Problems 6.3a

1 $o(\bar{4}) = 3$; $o(\bar{5}) = 6$; $o(\bar{6}) = 2$.

The following table shows that every element of $(\mathbb{Z}_7^*, \otimes)$ can be written as $(\bar{5})^r$, for some integer r.

r	1	2	3	4	5	6
$(\bar{5})^r$	$\bar{5}$	$\bar{4}$	$\bar{6}$	$\bar{2}$	$\bar{3}$	$\bar{1}$

The following table shows that it is not possible to express $\bar{3}, \bar{5}$ or $\bar{6}$ in the form of $(\bar{2})^r$, $r \in \mathbb{Z}$.

r	1	2	3	4	5	6	...
$(\bar{2})^r$	$\bar{2}$	$\bar{4}$	$\bar{1}$	$\bar{2}$	$\bar{4}$	$\bar{1}$	...

2 $o(\bar{2}) = 4$; $o(\bar{3}) = 8$; $o(\bar{4}) = 2$.

The following table shows that every element of $(\mathbb{Z}_8, \oplus)$ can be expressed as a multiple of $\bar{3}$.

r	1	2	3	4	5	6	7	8
$r(\bar{3})$	$\bar{3}$	$\bar{6}$	$\bar{1}$	$\bar{4}$	$\bar{7}$	$\bar{2}$	$\bar{5}$	$\bar{0}$

3 $o(\bar{1}) = 1$; $o(\bar{3}) = 4$; $o(\bar{7}) = 4$; $o(\bar{9}) = 2$.

4 The identity element of $(\mathbb{C}^*, \times)$ is $1 + 0i$.
Hence, z has *finite* order $m \Leftrightarrow z^m = 1 + 0i$, for some $m \in \mathbb{Z}^+$.
By DeMoivre's theorem, $z^m = r^m (\cos m\theta + i \sin m\theta)$, where $r \in \mathbb{R}^+$.
Also, $1 + 0i = 1 (\cos 2k\pi + i \sin 2k\pi)$, $k \in \mathbb{Z}$.
Hence $z^m = 1 + 0i \Leftrightarrow r^m = 1$ and $m\pi = 2k\pi$

$$\Leftrightarrow r = 1 \text{ and } \theta = \frac{2k\pi}{m}, \, m, k \in \mathbb{Z}.$$

5 (a) $ta = sa \Leftrightarrow t \equiv s(\text{mod } m)$; (b) $-(-a) = a$;
(c) $a + b = a + c \Rightarrow b = c$.

6 (a) In additive notation, $(ab)^{-1} = b^{-1}a^{-1}$ becomes
$$-(a + b) = (-b) + (-a).$$
But additive groups are all *abelian*, by convention.
Hence $(-b) + (-a) = (-a) + (-b)$.
(b) $a + x = b \Rightarrow x = (-a) + b = b + (-a)$ (see above);
$y + a = b \Rightarrow y = b + (-a)$.
Yes; $x = y = b + (-a)$ is often written $b - a$.

7 $(xy)^r = (xy)(xy) \ldots (xy)$ (r terms)
     $= (y^{-1}y)(xy)(xy) \ldots (xy)$
     $= y^{-1}(yx)(yx) \ldots (yx)y$, by associativity,
     $= y^{-1} (yx)^r y$.
Suppose $(yx)^r = e$. Then $(xy)^r = y^{-1} e y = e$.
But $(yx)^r = y (xy)^r y^{-1}$. Hence $(xy)^r = e \Rightarrow (yx)^r = y e y^{-1} = e$.
Thus $(yx)^r = e \Leftrightarrow (xy)^r = e$, and the result follows.

Problems 6.4a

1 $o(i) = o(-i) = 4$. Hence $(H, \times)$ is cyclic and both i and $-i$ are
generators by result 6.4.2. Since $o(1) = 1$ and $o(-1) = 2$, neither 1
nor -1 is a generator.

2 The matrix $I = \begin{pmatrix} 1 & 0 \\ 0 & 1 \end{pmatrix}$ is the identity element of K. Let A be any
matrix in K. Then $A^2 = I$ and hence the non-identity elements of K
each have order 2. Thus K is not cyclic, by result 6.4.2.

3 Since $o(f_1) = 1$, $o(f_2) = o(f_3) = o(f_5) = 2$ and $o(f_4) = o(f_6) = 3$,
$(M, \circ)$ is not cyclic, by result 6.4.2.

4 (a) Cyclic; generators: $\bar{2}, \bar{5}$.
Table to show that $(S, \otimes) = \langle \bar{2} \rangle$:

r	1	2	3	4	5	6
$(\bar{2})^r$	$\bar{2}$	$\bar{4}$	$\bar{8}$	$\bar{7}$	$\bar{5}$	$\bar{1}$

(b) Not cyclic; $o(\bar{5}) = o(\bar{7}) = o(\bar{11}) = 2$, $o(\bar{1}) = 1$.
(c) Cyclic; generators $\bar{3}, \bar{5}$.
Table to show that $(\mathbb{Z}_7^*, \otimes) = \langle \bar{3} \rangle$:

r	1	2	3	4	5	6
$(\bar{3})^r$	$\bar{3}$	$\bar{2}$	$\bar{6}$	$\bar{4}$	$\bar{5}$	$\bar{1}$

5 (a) $\bar{6}$; (b) $\bar{4}, \bar{8}$; (c) $\bar{3}, \bar{6}, \bar{9}$; (d) $\bar{2}, \bar{10}$; (e) $\bar{1}, \bar{5}, \bar{7}, \bar{11}$. Suppose
$o(\bar{a}) = 5$. Since $(\mathbb{Z}_{12}, \oplus) = \langle \bar{1} \rangle$, we can write $\bar{a} = a(\bar{1})$, where
$0 \le a < 12$. Then $5(\bar{a}) = \bar{0} \Rightarrow 5a(\bar{1}) = \bar{0} \Rightarrow 12|5a \Rightarrow a = 0$. But
$o(\bar{0}) = 1$. Contradiction.

6 (a) $\bar{1}, \bar{3}, \bar{7}, \bar{9}$; (b) $\bar{1}, \bar{2}, \bar{3}, \bar{4}, \bar{5}, \bar{6}$.

7 Suppose $M(2, \mathbb{Z}) = \left\langle \begin{pmatrix} a & b \\ c & d \end{pmatrix} \right\rangle$. Then every matrix of $M(2, \mathbb{Z})$ is of
the form $\begin{pmatrix} ra & rb \\ rc & rd \end{pmatrix}$, where $r, a, b, c, d \in \mathbb{Z}$.
But $\begin{pmatrix} 1 & 0 \\ 0 & 0 \end{pmatrix} \in M(2, \mathbb{Z})$. Hence $a = \pm 1$ and $b = c = d = 0$.
But $\begin{pmatrix} 0 & 1 \\ 0 & 0 \end{pmatrix} \in M(2, \mathbb{Z})$, giving a contradiction. Thus $M(2, \mathbb{Z})$ is not
cyclic.

8 Generalise the proof of example 6.4.3, 1 by writing g for 2 and g^{-1}
for $\frac{1}{2}$ throughout.

9 Suppose $(\mathbb{Q}^*, \times) = \left\langle \dfrac{a}{b} \right\rangle$, where $a \in \mathbb{Z}$, $b \in \mathbb{Z}^*$ and $\gcd(a, b) = 1$.

Then $\frac{1}{2} \in \mathbb{Q}^* \Rightarrow \frac{1}{2} = \left(\dfrac{a}{b} \right)^r$, for some $r \in \mathbb{Z} \Rightarrow a^r = 1 \Rightarrow a = \pm 1$,
which is clearly false, since, e.g., $\frac{2}{3} \in \mathbb{Q}^*$.

10 Let $(G, \times)$ be the group of n the complex roots of unity. Then by DeMoivre's theorem, $z^n = 1 \Leftrightarrow z = 1\left(\cos \dfrac{2k\pi}{n} + i \sin \dfrac{2k\pi}{n}\right)$

$= 1\left(\cos \dfrac{2\pi}{n} + \sin \dfrac{2\pi}{n}\right)^k$, for some $k \in \mathbb{Z}$.

Hence $(G, \otimes) = \langle z_1 \rangle$ where $z_1 = 1\left(\cos \dfrac{2\pi}{n} + i \sin \dfrac{2\pi}{n}\right)$.

Problems 6.5.1a

1 (a) $\begin{pmatrix} 1 & 2 & 3 & 4 & 5 \\ 1 & 5 & 4 & 3 & 2 \end{pmatrix}$; (b) $\begin{pmatrix} 1 & 2 & 3 & 4 & 5 \\ 5 & 4 & 3 & 2 & 1 \end{pmatrix}$;

(c) $\begin{pmatrix} 1 & 2 & 3 & 4 & 5 \\ 3 & 2 & 1 & 5 & 4 \end{pmatrix}$; (d) $\begin{pmatrix} 1 & 2 & 3 & 4 & 5 \\ 1 & 2 & 3 & 4 & 5 \end{pmatrix}$;

(e) 2; (f) $\begin{pmatrix} 1 & 2 & 3 & 4 & 5 \\ 3 & 4 & 5 & 1 & 2 \end{pmatrix}$;

(g) $\begin{pmatrix} 1 & 2 & 3 & 4 & 5 \\ 3 & 4 & 5 & 1 & 2 \end{pmatrix}$; (h) 5.

2 S_3:

$\circ$	ι	α_1	α_2	β_1	β_2	β_3
ι	ι	α_1	α_2	β_1	β_2	β_3
α_1	α_1	α_2	ι	β_3	β_1	β_2
α_2	α_2	ι	α_1	β_2	β_3	β_1
β_1	β_1	β_2	β_3	ι	α_1	α_2
β_2	β_2	β_3	β_1	α_2	ι	α_1
β_3	β_3	β_1	β_2	α_1	α_2	ι

3 $o(\iota) = 1$; $o(\alpha_1) = o(\alpha_2) = 3$; $o(\beta_1) = o(\beta_2) = o(\beta_3) = 2$.

4 (a) $\begin{pmatrix} 1 & 2 & 3 & 4 \\ 1 & 2 & 3 & 4 \end{pmatrix}$, $\begin{pmatrix} 1 & 2 & 3 & 4 \\ 1 & 3 & 4 & 2 \end{pmatrix}$, $\begin{pmatrix} 1 & 2 & 3 & 4 \\ 1 & 4 & 2 & 3 \end{pmatrix}$,

$\begin{pmatrix} 1 & 2 & 3 & 4 \\ 1 & 2 & 4 & 3 \end{pmatrix}$, $\begin{pmatrix} 1 & 2 & 3 & 4 \\ 1 & 4 & 3 & 2 \end{pmatrix}$, $\begin{pmatrix} 1 & 2 & 3 & 4 \\ 1 & 3 & 2 & 4 \end{pmatrix}$.

(b) The six permutations obtained by interchanging 1 and 2 in the second row of each of the permutations in (a) above; $(n-1)!$

5 $\alpha \circ \beta = \begin{pmatrix} 1 & 2 & 3 \\ 3 & 2 & 1 \end{pmatrix}$; Hence $(\alpha \circ \beta)^2 = \begin{pmatrix} 1 & 2 & 3 \\ 1 & 2 & 3 \end{pmatrix}$.

But $\alpha^2 = \begin{pmatrix} 1 & 2 & 3 \\ 1 & 2 & 3 \end{pmatrix}$, $\beta^2 = \begin{pmatrix} 1 & 2 & 3 \\ 2 & 3 & 1 \end{pmatrix}$; hence $\alpha^2 \circ \beta^2 = \begin{pmatrix} 1 & 2 & 3 \\ 2 & 3 & 1 \end{pmatrix}$.

Problems 6.5.2a

1 (a) $(1\ \ 3\ \ 4)(2\ \ 5)$;   (b) $(1\ \ 4\ \ 3\ \ 7)(2\ \ 6\ \ 5)$;
   (c) $(1\ \ 6)(2\ \ 4\ \ 7)(3\ \ 5)$;   (d) $(1\ \ 5\ \ 3\ \ 7\ \ 2\ \ 8\ \ 4)$.

2 (a) 6; (b) 12; (c) 6; (d) 7.

3 (a) 5; (b) 2; (c) 4; (d) 10; (e) 3.

4 (a) $(1\ \ 2\ \ 3\ \ 6\ \ 5)$, 5;   (b) $(1\ \ 4\ \ 2)(3\ \ 5)$, 6;
   (c) $(1\ \ 6)(2\ \ 4\ \ 3\ \ 5)$, 4;   (d) $(1\ \ 3\ \ 4\ \ 2\ \ 5\ \ 6)$; 6.

5 $\beta\gamma$: $x_1\ x_2 \ldots x_k\ y_1\ y_2 \ldots y_l$ $\gamma\beta$: $x_1\ x_2 \ldots x_k\ y_1\ y_2 \ldots y_l$
$\ x_1\ x_2 \ldots x_k\ y_2\ y_3 \ldots y_1$ $\ x_2\ x_3 \ldots x_1\ y_1\ y_2 \ldots y_l$
$\ x_2\ x_3 \ldots x_1\ y_2\ y_3 \ldots y_1$; $\ x_2\ x_3 \ldots x_1\ y_2\ y_3 \ldots y_1$.

6 $(1\ \ 2\ \ 3)$, $(1\ \ 3\ \ 2)$, $(1\ \ 2\ \ 4)$. $(1\ \ 4\ \ 2)$, $(1\ \ 3\ \ 4)$, $(1\ \ 4\ \ 3)$,
   $(2\ \ 3\ \ 4)$, $(2\ \ 4\ \ 3)$.

7 $(1\ \ 2\ \ 3\ \ 4\ \ 5\ \ 6)$; $(x_1 x_k)(x_1 x_{k-1}) \ldots (x_1 x_3)(x_1 x_2)$ or
   $(x_1 x_2)(x_2 x_3) \ldots (x_{k-1} x_k)$.

8 $(1\ \ 2\ \ 3)$, $(1\ \ 3\ \ 2)$, $(1\ \ 2)$, $(2\ \ 3)$, $(1\ \ 3)$.

9 $(1\ \ 5\ \ 4\ \ 3\ \ 2)$; $(x_1 x_k x_{k-1} \ldots x_3 x_2)$.

10 (a) $(1\ \ 4\ \ 3\ \ 2)$; $(1\ \ 2)(3\ \ 5\ \ 4)$ or $(3\ \ 5\ \ 4)(1\ \ 2)$;
   $(3\ \ 5\ \ 4)(1\ \ 3\ \ 2)$ (the order matters here because the cycles are
   not disjoint and do not commute), or $(1\ \ 5\ \ 4\ \ 3\ \ 2)$.

Problems 6.5.3a

1 (a) $(1\ \ 3\ \ 2\ \ 5) = (1\ \ 5)(1\ \ 2)(1\ \ 3)$, *odd*;
   (b) $(1\ \ 2\ \ 4)(3\ \ 5) = (1\ \ 4)(1\ \ 2)(3\ \ 5)$, *odd*;
   (c) $(1\ \ 2\ \ 4)(2\ \ 3\ \ 5) = (1\ \ 4)(1\ \ 2)(2\ \ 5)(2\ \ 3)$, *even*;
   (d) $(1\ \ 3\ \ 2\ \ 5)(2\ \ 4)(3\ \ 4\ \ 5)$
   $= (1\ \ 5)(1\ \ 2)(1\ \ 3)(2\ \ 4)(3\ \ 5)(3\ \ 4)$, *even*.

2 (a) Using the factorisation $(x_1 x_2 \ldots x_k) = (x_1 x_k) \ldots (x_1 x_3)(x_1 x_2)$ we
   see that a k-cycle can be expressed as the product of $k-1$ transpositions and result follows.
   (b) From (a) and result 6.5.6, we have for (i) $(1\ \ 2\ \ 3)$, or any other
   3-cycle; (ii) $(1\ \ 2)(3\ \ 4)$, or the product of any other pair of *disjoint*
   2-cycles; (iii) $(1\ \ 2)$, or any other transposition or any 4-cycle.
   Since this exhausts all the types of permutation of S_4, other than
   the identity, which is even, it is not possible to find an odd permutation with odd order in S_4. (See problems 6.5b. 3.)

3 (a) $(1\ \ 2)$; (b) $(1\ \ 2)(1\ \ 3)$; (c) $(2\ \ 3)(1\ \ 4)(1\ \ 3)(2\ \ 4)$;
   (d) $\beta_k \beta_{k-1} \ldots \beta_2 \beta_1$.

4 (a) and (b). Let $\alpha = \beta \circ \gamma$ and suppose that β has an expression as the product of r transpositions and γ as the product of s transpositions. Then clearly α can be expressed as the product of $r + s$ transpositions. By theorem 6.5.11, α is even $\Leftrightarrow r + s$ is even. Hence α is odd $\Leftrightarrow$ just one of β and γ is odd (again by theorem 6.5.11), proving (b). But in all other circumstances α is even, proving (a).

(c) Suppose $\alpha = \beta_1\beta_2 \ldots \beta_k$ is an expression for α as the product of k transpositions. Then $\alpha^{-1} = \beta_k \ldots \beta_2\beta_1$ is an expression for α^{-1}. Hence α^{-1} is even when α is even and odd when α is odd, by theorem 6.5.11.

Problems 6.6a

1 $\langle\alpha\rangle = \{\iota, \alpha_1, \alpha_2\}$; $\langle\beta_1\rangle = \{\iota, \beta_1\}$;
$\alpha_1 = (1 \quad 2 \quad 3)$, $o(\alpha_1) = 3$; $\beta_1 = (2 \quad 3)$, $o(\beta_1) = 2$.
The set is not closed: $\alpha_1\beta_1 = (1 \quad 2 \quad 3)(2 \quad 3) = (1 \quad 2) = \beta_3$, for example.

2 $\langle a \rangle = \{ra: r \in \mathbb{Z}\}$; $\langle 2 \rangle = \{2r: r \in \mathbb{Z}\} = \{0, \pm 2, \pm 4, \ldots\}$;
$\langle 3 \rangle = \{3r: r \in \mathbb{Z}\} = \{0, \pm 3, \pm 6, \ldots\}$.
(a) $\langle 2 \rangle \cap \langle 3 \rangle = \{0, \pm 6, \pm 12, \ldots\} = \{6r: r \in \mathbb{Z}\} = \langle 6 \rangle$.
(b) $\langle 4 \rangle = \{0, \pm 4, \pm 8, \pm 12, \ldots\}$;
$\langle 6 \rangle = \{0, \pm 6, \pm 12, \pm 18, \ldots\}$.
Hence $\langle 4 \rangle \cap \langle 6 \rangle = \{0, \pm 12, \pm 24, \ldots\} = \langle 12 \rangle$.
Proof: $x \in \langle 4 \rangle \cap \langle 6 \rangle \Rightarrow 4|x$ and $6|x$ and thus $12|x$ and $x \in \langle 12 \rangle$.
Conversely, if $y \in \langle 12 \rangle$ then $y = 12t$, for some $t \in \mathbb{Z}$.
Hence $4|y$ and $6|y$ and hence $y \in \langle 4 \rangle \cap \langle 6 \rangle = X$.
Thus $X \subseteq \langle 12 \rangle$ and $\langle 12 \rangle \subseteq X$ and hence $X = \langle 12 \rangle$.
(c) $l = \text{lcm}(a, b) = ab/\gcd(a, b)$. (See section 4.5.)

3 $\langle 2 \rangle \cup \langle 3 \rangle$ is not closed, since, for example, $2 \in \langle 2 \rangle$, $3 \in \langle 3 \rangle$ and hence $2, 3 \in \langle 2 \rangle \cup \langle 3 \rangle$. But $2 + 3 = 5 \notin \langle 2 \rangle \cup \langle 3 \rangle$.
(a) False; see counter-example above.
(b) False; suppose $H \subseteq K$ (or $K \subseteq H$). Then clearly $H \cup K = K$ (or H) and hence is a subgroup.
(c) True, from (a) and (b).

4 Let e be the identity of G. Then $e \in H$ and $e \in K$ and hence $e \notin H - K$. Hence $H - K$ has no identity element and cannot be a subgroup.

5 $\langle(1 \quad 2 \quad 3 \quad 4)\rangle = \{\iota, (1 \quad 2 \quad 3 \quad 4), (1 \quad 3)(2 \quad 4), (1 \quad 4 \quad 3 \quad 2)\}$.

6 $\langle\bar{0}\rangle = \{\bar{0}\}$; $\langle\bar{1}\rangle = \langle\bar{2}\rangle = \langle\bar{3}\rangle = \langle\bar{4}\rangle = \mathbb{Z}_5$. Hence the only subgroups of $\mathbb{Z}_5$ are $\{0\}$ and $\mathbb{Z}_5$ itself.

7 Let $\alpha = (1 \quad 2)$, $\beta = (3 \quad 4)$.

Then, $\alpha^2 = \beta^2 = (\alpha\beta)^2 = \iota$; hence each element is its own inverse. Closure is clear from the Cayley table:

$\circ$	ι	α	β	$\alpha\beta$
ι	ι	α	β	$\alpha\beta$
α	α	ι	$\alpha\beta$	β
β	β	$\alpha\beta$	ι	α
$\alpha\beta$	$\alpha\beta$	β	α	ι

8 Writing the elements $\alpha \in S_3$ in cycle form gives the following table for α^2 and α^3:

α	ι	(1 2 3)	(1 3 2)	(1 2)	(1 3)	(2 3)
α^2	ι	(1 3 2)	(1 2 3)	ι	ι	ι
α^3	ι	ι	ι	(1 2)	(1 3)	(2 3)

Hence $H = \{\iota, (1\ \ 2\ \ 3), (1\ \ 3\ \ 2)\}$ and it is easy to check this is the subgroup $\langle (1\ \ 2\ \ 3) \rangle$.
But $K = \{\iota, (1\ \ 2), (1\ \ 3), (2\ \ 3)\}$, which is not closed, since $(1\ \ 2)(1\ \ 3) = (1\ \ 3\ \ 2) \notin K$, for example. Hence K is not a subgroup.

9 (a) Let $P = \{\alpha \in S_n : \alpha(1) = 1\}$. Then:
    (i) P is non-empty, since for example $\iota \in P$.
    (ii) Let $\alpha, \beta \in P$. Then $\alpha\beta(1) = \alpha(\beta(1)) = \alpha(1) = 1$.
      Hence $\alpha, \beta \in P \Rightarrow \alpha\beta \in P$ and P is closed.
    (iii) $\alpha \in P \Rightarrow \alpha(1) = 1 \Rightarrow \alpha^{-1}(1) = 1 \Rightarrow \alpha^{-1} \in P$.
      Hence P is a subgroup of S_n.

  (b) Let $Q = \{\alpha \in S_n : \alpha(1) = 2\}$. Then $\iota \notin Q$. Thus Q is not a subgroup.

10 (a) Clearly g commutes with itself. Thus $g \in C_g$ and C_g is non-empty (alternatively, $eg = ge = g \Rightarrow e \in C_g$, where e is the identity of G).
  (b) Let $x, y \in C_g$. Then we know $xg = gx$ and $yg = gy$.
    Consider xy; we have $(xy)g = x(yg) = x(gy) = (xg)y = (gx)y = g(xy)$. Thus $xy \in C_g$.
  (c) Let $x \in C_g$. Then $xg = gx \Rightarrow x^{-1}(xg)x^{-1} = x^{-1}(gx)x^{-1}$
    $\Rightarrow (x^{-1}x)gx^{-1} = x^{-1}g(xx^{-1}) \Rightarrow gx^{-1} = x^{-1}g \Rightarrow x^{-1} \in C_g$.
    Thus C_g is a subgroup.

11 Let e be the identity of G. Then
    (a) H is non-empty, since $e = e^{-1}$ and hence $e \in H$.
    (b) Suppose $x, y \in H$. Then $x = x^{-1}$ and $y = y^{-1}$.
       Now $(xy)^{-1} = y^{-1}x^{-1} = yx = xy$, since G is abelian.
       Hence $xy \in H$.
    (c) Since $(x^{-1})^{-1} = x, x = x^{-1} \Rightarrow (x^{-1})^{-1} = x = x^{-1}$.
       Thus $x \in H \Rightarrow x^{-1} \in H$.
       Hence H is a subgroup.

12 $A_n = \{\iota, \ (1\ 2\ 3), \ (1\ 3\ 2), \ (1\ 2\ 4), \ (1\ 4\ 2), \ (1\ 3\ 4), \ (1\ 4\ 3),$
    $(2\ 3\ 4), \ (2\ 4\ 3), \ (1\ 2)(3\ 4), \ (1\ 3)(2\ 4), \ (1\ 4)(2\ 3)\}.$

Problems 6.7a

1

$\circ$	ι	ρ_1	ρ_2	σ_1	σ_2	σ_3
ι	ι	ρ_1	ρ_2	σ_1	σ_2	σ_3
ρ_1	ρ_1	ρ_2	ι	σ_3	σ_1	σ_2
ρ_2	ρ_2	ι	ρ_1	σ_2	σ_3	σ_1
σ_1	σ_1	σ_2	σ_3	ι	ρ_1	ρ_2
σ_2	σ_2	σ_3	σ_1	ρ_2	ι	ρ_1
σ_3	σ_3	σ_1	σ_2	ρ_1	ρ_2	ι

Note: The symmetry in the row opposite θ and the column under ϕ represents $\theta \circ \phi$. But remember that $\theta \circ \phi$ is calculated by first applying ϕ, then applying θ.

2 The Cayley table for the subset $H = \{\iota, \rho_1, \rho_2\}$ forms the top left-hand corner of the table in problem 1 above. From this we can see that (a) H is closed; (b) $\iota^{-1} = \iota$, $\rho_1^{-1} = \rho_2$, $\rho_2^{-1} = \rho_1$ and so H is also closed under inverses. Hence H is a subgroup. Let $K = \{\iota, \sigma_1, \sigma_2, \sigma_3\}$. Then K is not a subgroup, since it is not closed: $\sigma_1 \circ \sigma_2 = \rho_1$, for example.

3 $\beta_2 = \begin{pmatrix} 1 & 2 & 3 \\ 3 & 2 & 1 \end{pmatrix}; \quad \beta_3 = \begin{pmatrix} 1 & 2 & 3 \\ 2 & 1 & 3 \end{pmatrix}; \quad \alpha_2 = \begin{pmatrix} 1 & 2 & 3 \\ 3 & 1 & 2 \end{pmatrix}.$

4 (a) See Fig. 6.11.

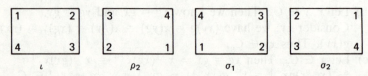

Fig. 6.11

(b)

∘	ι	ρ_2	σ_1	σ_2
ι	ι	ρ_2	σ_1	σ_2
ρ_2	ρ_2	ι	σ_2	σ_1
σ_1	σ_1	σ_2	ι	ρ_2
σ_2	σ_2	σ_1	ρ_2	ι

From the symmetry of the table about the main diagonal, we see that $\theta \circ \phi = \phi \circ \theta$, $\forall\,\theta, \phi$.

Hence the group is abelian.

(c) Let $M = \{\epsilon, \alpha_2, \beta_1, \beta_2\}$ where $\epsilon, \alpha_2, \beta_1, \beta_2$ are the permutations in S_4 corresponding to $\iota, \rho_2, \sigma_1, \alpha_2$ respectively.

Then $\epsilon = \begin{pmatrix} 1 & 2 & 3 & 4 \\ 1 & 2 & 3 & 4 \end{pmatrix}$; $\alpha_2 = \begin{pmatrix} 1 & 2 & 3 & 4 \\ 3 & 4 & 1 & 2 \end{pmatrix}$; $\beta_1 = \begin{pmatrix} 1 & 2 & 3 & 4 \\ 4 & 3 & 2 & 1 \end{pmatrix}$;

$\beta_2 = \begin{pmatrix} 1 & 2 & 3 & 4 \\ 2 & 1 & 4 & 3 \end{pmatrix}$.

∘	ι	α_2	β_1	β_2
ι	ι	α_2	β_1	β_2
α_2	α_2	ι	β_2	β_1
β_1	β_1	β_2	ι	α_2
β_2	β_2	β_1	α_2	ι

From the table, M is closed; also since each element is its own inverse, M is closed under inverses.

Thus M is a subgroup of S_4.

5 Four of the symmetries of a square (Fig. 6.12) are the symmetries ι, σ_1, σ_2 and ρ_2 defined in problem 4, above, for a rectangle. The other four are:

σ_3 flip about the axis AC;

σ_4 flip about the axis BD;

ρ_1 rotation through $\pi/2$ about X;

ρ_3 rotation through $3\pi/2$ about X, where X is the circumcentre.

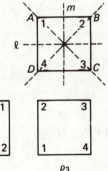

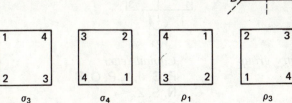

Fig. 6.12

∘	ι	ρ_1	ρ_2	ρ_3	σ_1	σ_2	σ_3	σ_4
ι	ι	ρ_1	ρ_2	ρ_3	σ_1	σ_2	σ_3	σ_4
ρ_1	ρ_1	ρ_2	ρ_3	ι	σ_3	σ_4	σ_2	σ_1
ρ_2	ρ_2	ρ_3	ι	ρ_1	σ_2	σ_1	σ_4	σ_3
ρ_3	ρ_3	ι	ρ_1	ρ_2	σ_4	σ_3	σ_1	σ_2
σ_1	σ_1	σ_4	σ_2	σ_3	ι	ρ_2	ρ_3	ρ_1
σ_2	σ_2	σ_3	σ_1	σ_4	ρ_2	ι	ρ_1	ρ_3
σ_3	σ_3	σ_1	σ_4	σ_2	ρ_1	ρ_3	ι	ρ_2
σ_4	σ_4	σ_2	σ_3	σ_1	ρ_3	ρ_1	ρ_2	ι

6 Sets (a) $\{\iota, \rho_1, \rho_2, \rho_3\}$ of rotations of the square,

(b) $\{\iota, \rho_2, \sigma_1, \sigma_2\}$ corresponding to the group of symmetries of the rectangle,

(c) $\{\iota, \rho_2, \sigma_3, \sigma_4\}$,

are all subgroups of the octic group, D_4.

7 For $\epsilon, \alpha_2, \beta_1, \beta_2$ corresponding to $\iota, \rho_2, \sigma_1, \sigma_2$ respectively, see problem 4 (c) above. Let $\beta_3, \beta_4, \alpha_1, \alpha_3$ correspond to $\sigma_3, \sigma_4, \rho_1, \rho_3$ respectively.

Then $\beta_3 = \begin{pmatrix} 1 & 2 & 3 & 4 \\ 1 & 4 & 3 & 2 \end{pmatrix}$; $\beta_4 = \begin{pmatrix} 1 & 2 & 3 & 4 \\ 3 & 2 & 1 & 4 \end{pmatrix}$;

$\alpha_1 = \begin{pmatrix} 1 & 2 & 3 & 4 \\ 2 & 3 & 4 & 1 \end{pmatrix}$; $\alpha_3 = \begin{pmatrix} 1 & 2 & 3 & 4 \\ 4 & 1 & 2 & 3 \end{pmatrix}$.

8 Subgroups of G: $\{\iota\}$, $\{\iota, \rho_2\}$, $\{\iota, \sigma_1\}$, $\{\iota, \sigma_2\}$, $\{\iota, \rho_2, \sigma_1, \sigma_2\}$.

From the table for G (see problem 4(b)), we can see that a subgroup of G containing any two of the non-identity elements of G must also contain the third, in order to be closed. Thus G contains no subgroups of order 3.

Imagine each of the capital letters of the alphabet inscribed in a rectangle (Fig. 6.13); test whether each of the symmetries of the rectangle is a symmetry of the resulting picture.

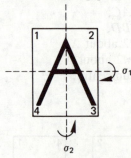

Fig. 6.13

Symmetry group	Capital letter
$\{\iota\}$	F, G, J, L, P, Q, R.
$\{\iota, \rho_2\}$	N, S, Z.
$\{\iota, \sigma_1\}$	B, C, D, E, K.
$\{\iota, \sigma_2\}$	A, M, T, U, V, W, Y.
$\{\iota, \rho_2, \sigma_1, \sigma_2\}$	H, I, O, X.

7 GROUPS: A FURTHER EXPLORATION

7.1 ISOMORPHISMS

In this section we look at groups which although they may have quite different elements and laws of combination nevertheless have the same *structure*.

Example 7.1.1

Consider the following groups of order 4:

(a) the subset $S = \{\bar{1}, \bar{3}, \bar{5}, \bar{7}\}$ of $\mathbb{Z}_8$, with respect to $\otimes$;

(b) the subset $T = \{\iota, \alpha_1 = (14)(23), \alpha_2 = (12)(34), \alpha_3 = (13)(24)\}$ of S_4, with respect to composition of mappings;

(c) the set $K =$
$$\left\{ I = \begin{pmatrix} 1 & 0 \\ 0 & 1 \end{pmatrix}, \quad A = \begin{pmatrix} -1 & 0 \\ 0 & 1 \end{pmatrix}, \quad B = \begin{pmatrix} 1 & 0 \\ 0 & -1 \end{pmatrix}, \quad C = \begin{pmatrix} -1 & 0 \\ 0 & -1 \end{pmatrix} \right\},$$
with respect to matrix multiplication.

Their Cayley tables are given by

(a)

$\otimes$	$\bar{1}$	$\bar{3}$	$\bar{5}$	$\bar{7}$
$\bar{1}$	$\bar{1}$	$\bar{3}$	$\bar{5}$	$\bar{7}$
$\bar{3}$	$\bar{3}$	$\bar{1}$	$\bar{7}$	$\bar{5}$
$\bar{5}$	$\bar{5}$	$\bar{7}$	$\bar{1}$	$\bar{3}$
$\bar{7}$	$\bar{7}$	$\bar{5}$	$\bar{3}$	$\bar{1}$

(b)

$\circ$	ι	α_1	α_2	α_3
ι	ι	α_1	α_2	α_3
α_1	α_1	ι	α_3	α_2
α_2	α_2	α_3	ι	α_1
α_3	α_3	α_2	α_1	ι

(c)

$\times$	I	A	B	C
I	I	A	B	C
A	A	I	C	B
B	B	C	I	A
C	C	B	A	I

Now suppose that in S, we substitute the symbol ι for $\bar{1}$, α_1 for $\bar{3}$, α_2 for $\bar{5}$ and α_3 for $\bar{7}$, then not only do we change the set S into the set T, but we *transform the Cayley table of S into the Cayley table of T*. Similarly we may transform the table for S into the table for K by the substitution I for $\bar{1}$, A for $\bar{3}$, B for $\bar{5}$ and C for $\bar{7}$. Thus each of the groups S, T and K may be transformed into either of the others by a suitable relabelling of its elements. So all three groups have the same algebraic structure.

Isomorphic groups

If two groups have exactly the same algebraic structure, so that one may be obtained from the other by a relabelling of its elements, then the groups are said to be *isomorphic*.

Condition for groups to be isomorphic

We now formalise this definition so that we know exactly what needs to be established to prove that two groups $(G, *)$ and $(H, \circ)$ are isomorphic. First, the 'relabelling' of the group elements can be done if and only if there is a bijective mapping ϕ, say, of G into H. Now consider when G and H are finite groups so that we can imagine a Cayley table for each. Any pair of elements $x, y \in G$ will be relabelled as $\phi(x)$, $\phi(y)$ when G is transformed into H. In the table for G the element in the row beside x and the column below y is $x * y$. So we must relabel $x * y$ as the element of H in the row beside $\phi(x)$ and the column below $\phi(y)$. Thus:

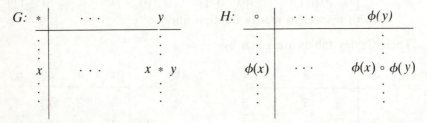

Fig. 7.1

Hence we need $\phi(x * y) = \phi(x) \circ \phi(y)$. If ϕ can be chosen to satisfy this condition for every pair of elements $x, y \in G$, then ϕ will transform the group G into the group H. This will also suffice to show that G and H have the same structure in the case when G and H are infinite, so we have the following definition:

Isomorphism

Let $(G, *)$, $(H, \circ)$ be two groups and $\phi\colon G \to H$ a mapping. Then ϕ is an *isomorphic mapping* or *isomorphism* of the groups $(G, *)$ and $(H, \circ)$ if:
(i) ϕ is bijective;
(ii) $\phi(x * y) = \phi(x) \circ \phi(y)$, for every pair of elements $x, y \in G$.
 If such a mapping ϕ exists, then the groups $(G, *)$ and $(H, \circ)$ are said to be *isomorphic*.

Notation

If groups G and H are isomorphic, we write

$$(G, *) \cong (H, \circ) \qquad \text{or simply} \qquad G \cong H.$$

Properties of isomorphic mappings

If there is a bijective mapping between groups $(G, *)$ and $(H, \circ)$, then there will normally be many such mappings (for example, there are 24 bijections of S into H in example 7.1.1). However, only a few (and possibly none) of these will satisfy condition (b) for an *isomorphism*. The following results establish properties of isomorphic mappings which follow from the definition. They give a very strong guide as to how an isomorphism between the groups (if it exists) should be defined. They can also be used to show that two groups cannot be isomorphic.

   In the following results let $\phi\colon (G, *) \to (H, \circ)$ be an *isomorphism*. Let e, f be the identity elements of G and H respectively.

RESULT 7.1.1 $\phi(e) = f$.

Proof   Let $x \in G$ be any element. Then in G

$$e * x = x.$$

Hence $\phi(e * x) = \phi(x)$.
But ϕ is an isomorphism and hence

$$\phi(e * x) = \phi(e) \circ \phi(x).$$

Thus $\phi(e) \circ \phi(x) = \phi(x)$.
But $f \circ \phi(x) = \phi(x)$ and hence $\phi(e) \circ \phi(x) = f \circ \phi(x)$.
$\Rightarrow$ $\phi(e) = f$, by the cancellation law (theorem 6.2.6). ∎

We illustrate this result in Fig. 7.2.

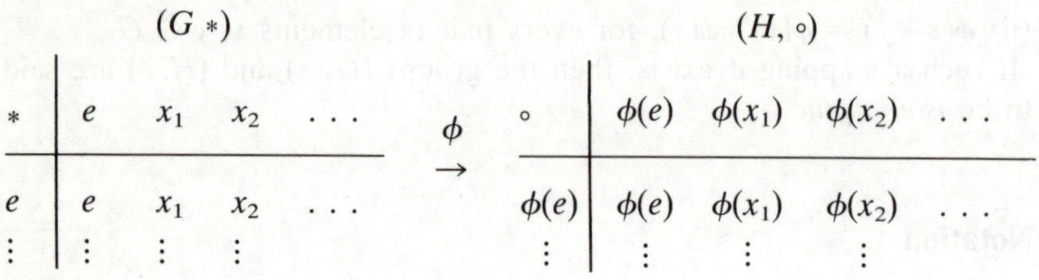

$(G *)$					ϕ	$(H, \circ)$			
$*$	e	x_1	x_2	$\ldots$	$\to$	$\circ$	$\phi(e)$	$\phi(x_1)$	$\phi(x_2)$ $\ldots$
e	e	x_1	x_2	$\ldots$		$\phi(e)$	$\phi(e)$	$\phi(x_1)$	$\phi(x_2)$ $\ldots$
$\vdots$	$\vdots$	$\vdots$	$\vdots$			$\vdots$	$\vdots$	$\vdots$	$\vdots$

Fig. 7.2

We have used the fact that because ϕ is an isomorphism, ϕ transforms the Cayley table of G into the Cayley table of H.

Since ϕ preserves the *structure* of G it seems natural that when ϕ maps x into y, then ϕ also maps the *inverse* of x into the *inverse* of y. We give a formal proof that this is indeed the case in result 7.1.2; the steps of the proof are as follows:

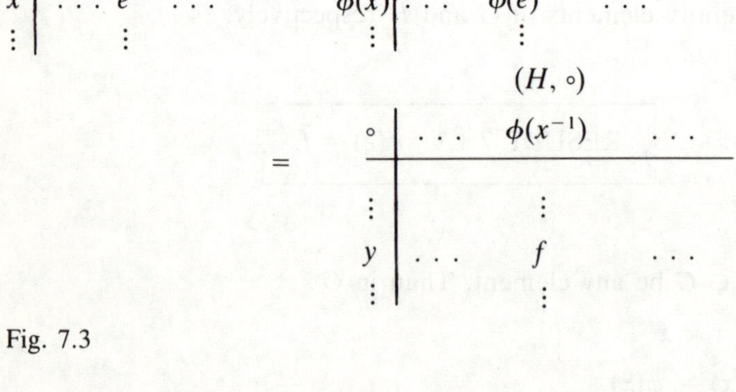

$(G, *)$				ϕ	$(H, \circ)$		
$*$	$\ldots x^{-1} \ldots$			$\to$	$\circ$	$\ldots$ $\phi(x^{-1})$ $\ldots$	
$\vdots$	$\vdots$				$\vdots$	$\vdots$	
x	$\ldots e \ldots$				$\phi(x)$	$\ldots$ $\phi(e)$ $\ldots$	
$\vdots$	$\vdots$				$\vdots$	$\vdots$	

$(H, \circ)$		
$\circ$	$\ldots$ $\phi(x^{-1})$ $\ldots$	
$=$		
$\vdots$	$\vdots$	
y	$\ldots$ f $\ldots$	
$\vdots$	$\vdots$	

Fig. 7.3

RESULT 7.1.2   Let $\phi(x) = y$. Then $\phi(x^{-1}) = y^{-1}$, for all $x \in G$.

Proof   In G, $x * x^{-1} = e$ and hence $\phi(x * x^{-1}) = \phi(e) = f$, by result 7.1.1.

But since ϕ is an isomorphism,

$$\phi(x * x^{-1}) = \phi(x) \circ \phi(x^{-1}) = y \circ \phi(x^{-1}).$$

Hence $y \circ \phi(x^{-1}) = f$.

Thus $\phi(x^{-1}) = y^{-1}$. ∎

It should seem equally natural that $\phi(x) = y \Rightarrow \phi(x^2) = y^2$.
Why this is so is illustrated in Fig. 7.4.

$(G, *)$					ϕ	$(H, \circ)$			
$*$	e	x	x^2	$\ldots$	$\rightarrow$	$\circ$	$\phi(e)$	$\phi(x)$	$\phi(x^2)$ $\ldots$
$\vdots$	$\vdots$	$\vdots$	$\vdots$			$\vdots$	$\vdots$	$\vdots$	$\vdots$
x	x	x^2	x^3 $\ldots$			$\phi(x)$	$\phi(x)$	$\phi(x^2)$	$\phi(x^3)$ $\ldots$
$\vdots$	$\vdots$	$\vdots$	$\vdots$			$\vdots$	$\vdots$	$\vdots$	$\vdots$

$(H, \circ)$

$=$	$\circ$	f	y $\ldots$
	$\vdots$	$\vdots$	$\vdots$
	y	y	$\phi(x^2) \ldots$
	$\vdots$	$\vdots$	$\vdots$

Fig. 7.4

Now once we have established that $\phi(x^2) = y^2$ we can extend the second table for $(H, \circ)$ to show that $\phi(x^3) = y^3$, and so on. This suggests that $\phi(x^n) = y^n$, for all positive integers n, and the words 'and so on' suggest a proof by induction.

RESULT 7.1.3   Let $\phi(x) = y$. Then $\phi(x^n) = y^n$, for all $x \in G$, $n \in \mathbb{Z}^+$.

Proof   We prove this result by induction on n. Let $P(n)$ be the statement to be proved.

   (i) Clearly $P(1)$ is true.
   (ii) Suppose $P(n)$ is true for $1 \leq n < k$, for some positive integer k. We want to show that this implies $P(k)$ is also true.

In G, $x^k = x^{k-1} * x$ so that $\phi(x^k) = \phi(x^{k-1} * x)$.
But, since ϕ is an isomorphism,

$$\phi(x^{k-1} * x) = \phi(x^{k-1}) \circ \phi(x).$$

By the induction hypothesis, $P(k-1)$ is true and hence

$$\phi(x^{k-1}) = y^{k-1}.$$

Thus $\phi(x^k) = y^{k-1} \circ y$
                       $= y^k.$

Hence $P(k)$ is true and so by induction $P(n)$ is true for all $n \in \mathbb{Z}^+$. ∎

We can easily extend result 7.1.3 to cover negative powers of x as well.

RESULT 7.1.4   Let $\phi(x) = y$. Then $\phi(x^m) = y^m$ for all $x \in G, m \in \mathbb{Z}$.

Proof   See problem 7.1a, 3. ∎

So once we have fixed on the image of x, then the image of *every power* of x is also determined. But how do we decide upon the image of x in the first place? The following results give the possible candidates for $\phi(x)$.

RESULT 7.1.5   Let $x \in G$ be an element of finite order n. Then $\phi(x)$ is also an element of order n.

Proof   In G, $x^n = e$ and hence $\phi(x^n) = \phi(e) = f$, by result 7.1.1. Let $\phi(x) = y$. Then $\phi(x^n) = y^n$, by result 7.1.3. Hence

$$y^n = f.$$

To show that y has order n we must show that n is the *least* positive integer for which $y^n = f$. So suppose that $y^k = f$ for some k, $1 \leq k < n$.

But again by result 7.1.3, this implies $\phi(x^k) = f$ and hence $x^k = e$, since ϕ is bijective. But n is the least integer for which $x^n = f$. Thus no such k exists and y also has order n. ∎

The method of the last part of the proof can be used to show the following:

RESULT 7.1.6 Let $x \in G$ be an element of infinite order. Then $\phi(x)$ has infinite order in H.

Proof See problems 7.1b, 6. ∎

The proof of the following theorem illustrates how these ideas are used in constructing an isomorphism.

THEOREM 7.1.7

(a) Any cyclic group of finite order n is isomorphic to $(\mathbb{Z}_n, \oplus)$.
(b) Any infinite cyclic group is isomorphic to $(\mathbb{Z}, +)$.

Proof (a) Let $(G, \circ)$ be a cyclic group of order n. Then by definition, G must contain a generator x of order n. Then

$$G = \langle x \rangle = \{e, x, x^2, \ldots, x^{n-1}\},$$

where $x^n = e$.

Now $(\mathbb{Z}_n, \oplus)$ is also a cyclic group, generated by $\bar{1}$. We seek to define a bijective mapping $\phi: G \to \mathbb{Z}_n$ in such a way that ϕ is an isomorphism of $(G, \circ)$ and $(\mathbb{Z}_n, \oplus)$. From result 7.1.1, we must define $\phi(e) = \bar{0}$. From result 7.1.5, $\phi(x)$ must be an element of order n. Thus we can try defining $\phi(x) = \bar{1}$. Then by result 7.1.3, $\phi(x^r)$ is the rth 'power' of $\bar{1} = \bar{1} \oplus \bar{1} \oplus \ldots \oplus \bar{1} = \bar{r}$, for all r, $1 \leq r \leq n-1$. This completely defines a mapping $\phi: G \to \mathbb{Z}_n$ and clearly it is bijective.

To show that ϕ is an isomorphism, we must also show that given $y, z \in G$, then $\phi(yz) = \phi(y) \oplus \phi(z)$. Now we can find s, t such that $y = x^s$ and $z = x^t$.

Hence $yz = x^{s+t}$

$\Rightarrow \phi(yz) = \overline{s+t}$

$= \bar{s} \oplus \bar{t}$

$= \phi(y) \oplus \phi(z)$.

Thus ϕ is an isomorphism and $(G, \circ) \cong (\mathbb{Z}_n, \oplus)$

(b) Now suppose G is an infinite cyclic group with generator x. Then $G = \{x^r : r \in \mathbb{Z}\}$ and $(\mathbb{Z}, +)$ is generated by the element 1. Define $\phi: G \to \mathbb{Z}$ by

$$\phi(x^r) = r, \ r \in \mathbb{Z}.$$

Then clearly ϕ is bijective. The proof that ϕ is an isomorphism is completed by the same method as in (a) above. ■

Examples 7.1.2

1. Show that the group K of exercise 7.1.1 is not isomorphic to $(\mathbb{Z}_4, \oplus)$.

Solution The following tables show the *orders* of the elements in each group.

K: element	I	A	B	C		$(\mathbb{Z}_4, \oplus)$: element	$\bar{0}$	$\bar{1}$	$\bar{2}$	$\bar{3}$
order	1	2	2	2		order	1	4	2	4

By result 7.1.5 an isomorphic mapping $\phi: K \to \mathbb{Z}_4$ is impossible since K contains no element of order 4 to map onto $\bar{1}$ or $\bar{3}$.

2. Show that the groups $(\mathbb{R}^+, \times)$ and $(\mathbb{Q}^+, \times)$ are not isomorphic.

Solution If you know a little analysis you can prove this by showing that there is no bijective mapping from $\mathbb{R}^+$ onto $\mathbb{Q}^+$. Alternatively, we can look for a property of multiplication in $\mathbb{R}^+$ which is not true in $\mathbb{Q}^+$. One such is the existence of the square root of every number in $\mathbb{R}^+$, whereas not every number in $\mathbb{Q}^+$ has a square root in $\mathbb{Q}^+$. This idea gives the following proof by contradiction. Suppose there exists an isomorphism $\phi: (\mathbb{R}^+, \times) \to (\mathbb{Q}^+, \times)$. Then since ϕ is a bijection, we can find $a \in \mathbb{R}^+$ such that $\phi(a) = 2$. In $\mathbb{R}^+$, we can find x such that $x^2 = a$. Let $\phi(x) = y \in \mathbb{Q}^+$. Then $\phi(x^2) = y^2$, by result 7.1.3. But $\phi(x^2) = \phi(a) = 2$. Thus $y^2 = 2$, where $y \in \mathbb{Q}^+$, which is impossible (see theorem 4.5.4).

Isomorphism as an equivalence relation

We have said that two groups are isomorphic if they have exactly the same structure and so it should not be surprising that we can prove (see problem 7.1a, 13) that isomorphism is an *equivalence relation* on a set of groups. Each equivalence class is a set of groups which are all isomorphic to one another, that is, they all have the same algebraic structure. Theorem 7.1.7 tells us that for each each $n \in \mathbb{Z}^+$, all the cyclic groups of order n form an equivalence class; all infinite cyclic groups also form an equivalence class. The groups, S, T and K, of exercise 1.1 belong to the same equivalence class but example 7.1.2, 1, shows that this is not the same as the class of cyclic groups of order 4.

Isomorphism is a powerful idea because it allows us to consider just the structurally different *types* of group which can occur, rather than individual groups.

Klein 4-group

The isomorphism class to which the groups S, T and K of example 7.1.1 belong is known as the class of *Klein 4-groups*. By a suitable relabelling of the elements, the table for any Klein 4-group can be transformed into the table for the group V given below.

$V:$ $\circ$	e	a	b	c
e	e	a	b	c
a	a	e	c	b
b	b	c	e	a
c	c	b	a	e

The symbol V comes from the German word *Viergruppe*. In section 7.4 we shall show that any group of order 4 is isomorphic to either V or $(\mathbb{Z}_4, \oplus)$.

Cayley's theorem

In result 6.2.8 we showed that every row and every column of the Cayley table of a finite group G is a permutation of the elements of G.

THEOREM 7.1.8 (Cayley)   Every group is isomorphic to a permutation group.

Before proving this general statement of the theorem, we look first at the case where G is a finite group of order n. Let $G = \{g_1, g_2, \ldots, g_n\}$ and consider the row opposite an arbitrary element g in the Cayley table of G. We know from result 6.2.8 that this row is a permutation of the symbols $g_1, g_2, \ldots, g_n$.

Hence we can define a permutation $\alpha_g \in S_n$, the symmetric group on n symbols, by

$$\alpha_g = \begin{pmatrix} g_1 & g_2 & \cdots & g_n \\ gg_1 & gg_2 & \cdots & gg_n \end{pmatrix}.$$

$\circ$	g_1	g_2	$\cdots$	g_r	$\cdots$	g_n
$\vdots$	$\vdots$	$\vdots$		$\vdots$		$\vdots$
g	gg_1	gg_2	$\cdots$	gg_r	$\cdots$	gg_n
$\vdots$	$\vdots$	$\vdots$		$\vdots$		$\vdots$

Writing α_g more succinctly in mapping notation, we have

$$\alpha_g: x \mapsto gx, \ \forall\, x \in G.$$

Thus we can define a permutation α_g corresponding to every element $g \in G$. This gives us a set $H = \{\alpha_g : g \in G\}$ where H is a subset of S_n. To prove Cayley's theorem we need to show (i) H is a group and (ii) $G \cong H$. Now it so happens that nowhere in the proof of (i) and (ii) do we need to use the fact that G and H are finite. Also, we can show that the mapping $\alpha_g: x \mapsto gx, \ \forall\, x \in G$, is a permutation of G even in the case when g is infinite. The point, therefore, in looking at the finite case first was to show how the permutation α_g arises in a natural way from the group table. In the proof given below we do not assume that G is finite.

Proof of Theorem 7.1.8 Let G be a group and let $g \in G$ be any element. Define a mapping $\alpha_g: G \mapsto G$ by $\alpha_g: x \mapsto gx, \ \forall\, x \in G$. We shall show (a) α_g is a permutation of G; (b) the set $H = \{\alpha_g : g \in G\}$ is a group; (c) $G \cong H$.

(a) A permutation of G is a bijection of G onto G and so we must show that α_g is *surjective* and *injective*.

Let $y \in G$ be any element. Then by theorem 6.2.7 we can find $x \in G$ such that $gx = y$. For this element x, $\alpha_g(x) = gx = y$ and hence α_g is *surjective*. Now suppose that $\alpha_g(x_1) = \alpha_g(x_2)$. Then $gx_1 = gx_2 \Rightarrow x_1 = x_2$ by theorem 6.2.7. Hence α_g is *injective* and so α_g is a permutation of G.

(b) Let S be the group of all permutations on the set G. We know from (a) that H is a subset of S, so we need to show that H satisfies the conditions to be a *subgroup* of S. From theorem 6.6.2 we need to show that H is closed and that each element of H has an inverse in H. Suppose $\alpha_h, \alpha_g \in H$, where $h, g \in G$. We want to show $\alpha_h \circ \alpha_g \in H$. Now $\alpha_h \circ \alpha_g(x) = \alpha_h(\alpha_g(x)) = \alpha_h(gx) = h(gx) = (hg)x$, the last step following from the associative law in G. Now G is closed, so $hg \in G$. Let $hg = j$. Then there is a permutation $\alpha_j \in H$ defined by $\alpha_j(x) = jx = (hg)x$. Hence $\alpha_h \circ \alpha_g = \alpha_j \in H$. Thus H is closed.

Now consider any permutation $\alpha_g \in H$. Since G is a group, the element $g \in G$ has an inverse $g^{-1} \in G$, where $g^{-1}g = e$, the identity element of G. Corresponding to g^{-1}, there is a permutation $\alpha_{g^{-1}}$ in H defined by $\alpha_{g^{-1}}(x) = g^{-1}x$.

Hence $\alpha_{g^{-1}} \circ \alpha_g(x) = (g^{-1}g)x = ex = x$.

Thus $\alpha_{g^{-1}} \circ \alpha_g = \iota$, the identity permutation on G. Hence $\alpha_{g^{-1}}$ is the inverse permutation of α_g.

This concludes the proof that H is a subgroup of S.

(c) To show $G \cong H$ we must start by defining a bijective mapping of G onto H. The most natural choice of such a mapping ϕ is to match every element in G with the permutation it defines, that is let ϕ be given by

$$\phi: g \mapsto \alpha_g.$$

Since each permutation in H arises from an element of G, ϕ is *surjective*. If $\phi(g) = \phi(h)$, then $\alpha_g = \alpha_h$ and hence $gx = hx$, $\forall\ x \in G$, giving $g = h$. Thus ϕ is also *injective* and hence *bijective*.

Finally we must show that for all $h, g \in G$,

$$\phi(hg) = \phi(h) \circ \phi(g).$$

We do this by computing both sides.
Let $hg = j$. Then $\phi(hg) = \phi(j) = \alpha_j$.
But $\phi(h) \circ \phi(g) = \alpha_h \circ \alpha_g = \alpha_j$, as in proof of (b) above.
Hence ϕ is an isomorphism and $G \cong H$. This concludes the proof of the theorem. ■

The following example illustrates the theorem.

Example 7.1.3

Let G be group S in example 7.1.1. Let α_k be the permutation corresponding to $\bar{k}$. Then by selecting the row corresponding to each element in turn, we have

$$S: \quad
\begin{array}{c|cccc}
\otimes & \bar{1} & \bar{3} & \bar{5} & \bar{7} \\
\hline
\bar{1} & \bar{1} & \bar{3} & \bar{5} & \bar{7} \\
\bar{3} & \bar{3} & \bar{1} & \bar{7} & \bar{5} \\
\bar{5} & \bar{5} & \bar{7} & \bar{1} & \bar{3} \\
\bar{7} & \bar{7} & \bar{5} & \bar{3} & \bar{1}
\end{array}$$

$$\alpha_1 = \begin{pmatrix} \bar{1} & \bar{3} & \bar{5} & \bar{7} \\ \bar{1} & \bar{3} & \bar{5} & \bar{7} \end{pmatrix}; \quad \alpha_3 = \begin{pmatrix} \bar{1} & \bar{3} & \bar{5} & \bar{7} \\ \bar{3} & \bar{1} & \bar{7} & \bar{5} \end{pmatrix};$$

$$\alpha_5 = \begin{pmatrix} \overline{1} & \overline{3} & \overline{5} & \overline{7} \\ \overline{5} & \overline{7} & \overline{1} & \overline{3} \end{pmatrix}; \quad \alpha_7 = \begin{pmatrix} \overline{1} & \overline{3} & \overline{5} & \overline{7} \\ \overline{7} & \overline{5} & \overline{3} & \overline{1} \end{pmatrix}.$$

Combining the permutations in the usual way gives the following table for H:

H:	$\circ$	α_1	α_3	α_5	α_7
	α_1	α_1	α_3	α_5	α_7
	α_3	α_3	α_1	α_7	α_5
	α_5	α_5	α_7	α_1	α_3
	α_7	α_7	α_5	α_3	α_1

It is easy to see that the mapping $\phi: \overline{k} \to \alpha_k$ will transform the table for S into the table for H. Thus $S \cong H$.

PROBLEMS 7.1a

1 How many distinct isomorphic mappings are there of the group $(S, \otimes)$ into the group $(K, \times)$ of example 7.1.1?

2 For each of the following groups, give an isomorphic mapping into either V or $(\mathbb{Z}_4, \oplus)$, as appropriate.

(a) the subset $\{\overline{1}, \overline{3}, \overline{7}, \overline{9}\}$ of $\mathbb{Z}_{10}$ with respect to $\otimes$;
(b) the subset $\{1, -1, i, -i\}$ of $(\mathbb{C}^*, \times)$;
(c) the subset $\{\overline{1}, \overline{5}, \overline{7}, \overline{11}\}$ of $\mathbb{Z}_{12}$ with respect to $\otimes$.

3 Assuming results 7.1.1 to 7.1.3, prove result 7.1.4.

4 Let $G = \langle a \rangle$ be a cyclic group of order 6. Complete the table below so that the mappings $\phi_i: (G, \circ) \to (\mathbb{Z}_6, \oplus)$, $i = 1, 2$, are both isomorphisms. Is it possible to define any other isomorphic mapping of $(G, \circ)$ into $(\mathbb{Z}_6, \oplus)$? Explain.

x	e	a	a^2	a^3	a^4	a^5
$\phi_1(x)$	.	$\overline{1}$	.	.	.	.
$\phi_2(x)$	.	$\overline{5}$	.	.	.	.

5 An isomorphic mapping of a group G onto itself is called an *automorphism*.
Show that there are two automorphisms of $(\mathbb{Z}_6, \oplus)$.
Tabulate all the automorphisms of

(a) $(\mathbb{Z}_3, \oplus)$;   (b) $(\mathbb{Z}_4, \oplus)$.

6 Complete the table on the next page so that each mapping $\phi_i: S_3 \to S_3$ is an automorphism, $i = 1, 2, 3$.

α	ι	(1 2 3)	(1 3 2)	(2 3)	(1 3)	(1 2)
$\phi_1(\alpha)$	.	(1 2 3)	.		(1 3)	.	.
$\phi_2(\alpha)$	.	(1 2 3)	.		(1 2)	.	.
$\phi_3(\alpha)$	.	(1 3 2)	.		(1 2)	.	.

7 How many automorphisms are there of $(\mathbb{Z}, +)$?

8 Is the mapping $\phi: \mathbb{R}^* \to \mathbb{R}^*$ defined by $\phi(x) = 5x$ an automorphism of $(\mathbb{R}^*, \times)$? Justify your answer.

9 Prove that $(\mathbb{C}^*, \times)$ and $(\mathbb{R}^*, \times)$ are not isomorphic.

10 Prove that the mapping $\phi: \mathbb{R} \to \mathbb{R}^+$ defined by $\phi(x) = 2^x$ is an isomorphism of the groups $(\mathbb{R}, +)$ and $(\mathbb{R}^+, \times)$.

11 The set $W = \left\{ \begin{bmatrix} x & -x \\ 0 & 0 \end{bmatrix} : x \in \mathbb{Q}^* \right\}$ is a group with respect to matrix multiplication (see problems 6.1.1a, 4). Prove that $(W, \times) \cong (\mathbb{Q}^*, \times)$.

12 Is the following statement true or false? Give a proof or counter-example as appropriate.

No abelian group can be isomorphic to a non-abelian group.

13 Let G_1, G_2, G_3 be the three groups and suppose that $\theta: G_1 \to G_2$ and $\phi: G_2 \to G_3$ are isomorphisms. Prove that

(a) $\theta^{-1}: G_2 \to G_1$ is an isomorphism;
(b) $\phi \circ \theta: G_1 \to G_3$ is an isomorphism;
(c) every group is isomorphic to itself.

Deduce that *isomorphism* is an equivalence relation on a set of groups.

14 Let $\phi: G \to G'$ be an isomorphism. Let H be a subgroup of G. Prove that the set $\phi(H) = \{\phi(h): h \in H\}$ is a subgroup of G'.
What is the connection between the orders of H and $\phi(H)$? Justify your answer.

PROBLEMS 7.1b

1 The subset $H = \{\bar{2}, \bar{4}, \bar{6}, \bar{8}\}$ of $\mathbb{Z}_{10}$ is a group with respect to $\oplus$ (see problem 6.1b, 2). Find an isomorphic mapping of $(H, \oplus)$ onto $(\mathbb{Z}_4, \oplus)$.

2 Show that the groups S_3 and $(\mathbb{Z}_6, \oplus)$ are *not* isomorphic.

3 For each of the following groups, given an isomorphism into either S_3 or $(\mathbb{Z}_6, \oplus)$ as appropriate.

(a) $(\mathbb{Z}_7^*, \otimes)$;   (b) D_3;

(c) the group $(M, \circ)$ of mappings defined in problem 6.1.2a, 10;

(d) the group of complex sixth roots of unity, $\{z \in \mathbb{C}^*: z^6 = 1\}$, with respect to multiplication.

4 Let $G = \langle a \rangle$ be a cyclic group of order 5. Complete the following table so that each of the mappings $\phi_i: G \to (\mathbb{Z}_5, \oplus)$ is an isomorphism.

x	e	a	a^2	a^3	a^4
$\phi_1(x)$	.	a	.	.	.
$\phi_2(x)$	.	a^2	.	.	.
$\phi_3(x)$	.	a^3	.	.	.
$\phi_4(x)$	.	a^4	.	.	.

5 How many automorphisms are there of $(\mathbb{Z}_5, \oplus)$?

Show that there are just four automorphisms of $(\mathbb{Z}_8, \oplus)$.

Find the number of automorphisms of $(\mathbb{Z}_p, \oplus)$ where p is a prime. What can you say about the number of automorphisms of $(\mathbb{Z}_n, \oplus)$ when n is not necessarily a prime?

6 Prove result 7.1.6.

7 Prove that $(5\mathbb{Z}, +) \cong (7\mathbb{Z}, +)$.

8 Show that $(\mathbb{R}^*, \times)$ and $(\mathbb{Q}^*, \times)$ are not isomorphic by considering the existence of cube roots.

9 $S = \{2^n: n \in \mathbb{Z}\}$ is a group with respect to multiplication (see problem 6.1.1a, 1). Prove that $(S, \times) \cong (\mathbb{Z}, \oplus)$.

10 The set $W = \left\{ \begin{bmatrix} x & -x \\ -x & x \end{bmatrix} : x \in \mathbb{R}^* \right\}$ is a group with respect to matrix multiplication (see problem 6.1b, 6). Prove that $(W, \times) \cong (\mathbb{R}^*, \times)$.

11 Show that the set of matrices $M: \left\{ \begin{bmatrix} \cos x & \sin x \\ -\sin x & \cos x \end{bmatrix} : x \in \mathbb{R} \right\}$ is a subgroup of $GL(2, \mathbb{R})$. Let U be the subgroup of $(\mathbb{C}^*, \times)$ defined by $U = \{\cos x + i \sin x: x \in \mathbb{R}\}$. Prove that $(M, \times) \cong (U, \times)$.

12 Let $\circ$ be the binary operation defined on $\mathbb{Z}$ by $a \circ b = a + b + 1$. Then $(\mathbb{Z}, \circ)$ is a group (see problem 6.1.1a, 3).

Prove that $(\mathbb{Z}, \circ) \cong (\mathbb{Z}, +)$.

13 Let G be a group and $\phi: G \to G$ be the mapping defined by $\phi(x) = x^{-1}$.

Explain why ϕ is bijective and prove that ϕ is an isomorphism if and only if G is abelian.

14 Let G be a group and $a \in G$. The mapping $\phi_a: G \to G$ is defined by $\phi_a(x) = a^{-1}xa$, for all $x \in G$. Prove that ϕ_a is an automorphism of G. (A mapping defined in this way is called an *inner automorphism* of G.)

What can you say about ϕ_a if G is abelian?

Tabulate the inner automorphism ϕ_a of S_3 where $a = (1 \quad 2)$.

15 Find a subgroup of S_4 isomorphic to S_3.

7.2 COSETS AND LAGRANGE'S THEOREM

You will probably have noticed that in every case you have so far met it has been true that the order of a subgroup divides exactly the order of the group. In this section, we prove that this must always be so. To this end, we show how any subgroup can be used to *partition* the elements of a group into subsets called *cosets*, each containing the same number of elements.

Cosets

Let H be a subgroup of a group G and let $g \in G$ be any element. Then the subset $gH = \{gh: h \in H\}$ is called the *left coset* of H in G determined by the element g. Similarly, the set $Hg = \{hg: h \in H\}$ is called the *right coset* of H in G determined by g.

Examples 7.2.1

1. Consider the subgroup $H = \{\iota, (1 \quad 2 \quad 3), (1 \quad 3 \quad 2)\}$ of the subgroup S_3. The complete set of *left cosets* of H in S_3 are:

$$\iota H = \{\iota^2, \iota(1 \quad 2 \quad 3), \iota(1 \quad 3 \quad 2)\}$$
$$= \{\iota, (1 \quad 2 \quad 3), (1 \quad 3 \quad 2)\} = X_1;$$
$$(1 \quad 2)H = \{(1 \quad 2)\iota, (1 \quad 2)(1 \quad 2 \quad 3), (1 \quad 2)(1 \quad 3 \quad 2)\}$$
$$= \{(1 \quad 2), (2 \quad 3), (1 \quad 3)\} = X_2;$$
$$(2 \quad 3)H = \{(2 \quad 3)\iota, (2 \quad 3)(1 \quad 2 \quad 3), (2 \quad 3)(1 \quad 3 \quad 2)\}$$
$$= \{(2 \quad 3), (1 \quad 3), (1 \quad 2)\} = X_2;$$
$$(1 \quad 3)H = \{(1 \quad 3)\iota, (1 \quad 3)(1 \quad 2 \quad 3), (1 \quad 3)(1 \quad 3 \quad 2)\}$$
$$= \{(1 \quad 3), (1 \quad 2), (2 \quad 3)\} = X_2;$$
$$(1 \quad 2 \quad 3)H = \{(1 \quad 2 \quad 3)\iota, (1 \quad 2 \quad 3)^2, (1 \quad 2 \quad 3)(1 \quad 3 \quad 2)\}$$
$$= \{(1 \quad 2 \quad 3), (1 \quad 3 \quad 2), \iota\} = X_1;$$
$$(1 \quad 3 \quad 2)H = \{(1 \quad 3 \quad 2)\iota, (1 \quad 3 \quad 2)(1 \quad 2 \quad 3), (1 \quad 3 \quad 2)^2\}$$
$$= \{(1 \quad 3 \quad 2), \iota, (1 \quad 2 \quad 3)\} = X_1.$$

Notice that:

(a) only two of the cosets are distinct as *sets*; we have called these X_1 and X_2.
(b) the set $\Sigma = \{X_1, X_2\}$ partitions the six elements of S_3 into two subsets each containing $|H| = 3$ elements.

2. Consider the subgroup $H = \{\bar{0}, \bar{3}\}$ of $(\mathbb{Z}_6, \oplus)$. For an additive group, the *left cosets* are defined by

$$g + H = \{g + h : h \in H\}.$$

Hence the left cosets of H in $(\mathbb{Z}_6, \oplus)$ are:

$$\bar{0} + H = \{\bar{0}, \bar{3}\} = X_1; \quad \bar{3} + H = \{\bar{3}, \bar{0}\} = X_1$$
$$\bar{1} + H = \{\bar{1}, \bar{4}\} = X_2; \quad \bar{4} + H = \{\bar{4}, \bar{1}\} = X_2$$
$$\bar{2} + H = \{\bar{2}, \bar{5}\} = X_3; \quad \bar{5} + H = \{\bar{5}, \bar{2}\} = X_3.$$

This time we have three distinct cosets, X_1, X_2, X_3 and $\Sigma = \{X_1, X_2, X_3\}$ partitions $\mathbb{Z}_6$ into three subsets each containing $|H| = 2$ elements.

Our next task is to prove that for any group G the set of distinct left cosets of a subgroup H of G partition the elements of G in the way described above in examples 7.2.1. Now if the left cosets of H partition G, then it must be possible to define an *equivalence relation* on the elements of G in such a way that the cosets are the *equivalence classes* (see section 2.3). We now prove that this can always be done.

THEOREM 7.2.1 Let G be a group and H a subgroup of G. Define a relation $\sim$ on G by $x \sim y \Leftrightarrow x \in yH$. Then $\sim$ is an equivalence relation on G and the left cosets of H in G are the equivalence classes.

Proof We need to show that $\sim$ is (i) reflexive, (ii) symmetric and (iii) transitive.

(i) Let e denote the identity element of G. Then $e \in H$. Hence given any $x \in G$, $x = xe \in xH$. Thus $x \sim x$, $\forall x \in G$, and so $\sim$ is reflexive.
(ii) Let $x, y \in G$ be any elements such that $x \sim y$. Then $x \in yH$ and hence there exists $h \in H$ such that $x = yh$. Now $x = yh \Rightarrow y = xh^{-1}$. But $h \in H \Rightarrow h^{-1} \in H$ (subgroup property) and hence $y \in xH$. Thus $y \sim x$ and $\sim$ is symmetric.
(iii) Let $x, y, z \in G$ be any elements such that $x \sim y$ and $y \sim z$. Then $x \in yH$, $y \in zH$ and thus there exist elements $h, k \in H$ such that $x = yh$ and $y = zk$, giving $x = (zk)h = z(kh)$. But $k, h \in H \Rightarrow kh \in H$ (subgroup property) and hence $x \in zH$. Thus $x \sim z$ and $\sim$ is transitive.

This establishes that $\sim$ is an equivalence relation on G.

For any $x \in G$, the equivalence class $\bar{x}$ of x is defined by

$$\bar{x} = \{y \in G: y \sim x\}$$
$$= \{y \in G: y \in xH\}.$$

Thus the left coset xH is the equivalence class of x under the relation $\sim$, concluding the proof. ∎

COROLLARY 7.2.2   The set Σ of distinct left costs of H in G partitions G.

Theorem 2.1 holds for both finite and infinite groups. However a most important application of this result is in the case when G is finite because then it implies the following theorem, due to the famous French mathematician Joseph Louis Lagrange (1736–1813). It is one of the fundamental theorems of group theory.

THEOREM 7.2.3 (Lagrange)   Let G be a group of finite order n. Then the order of every subgroup H of G is a divisor of n.

Proof   Let H be a subgroup of G and suppose $|H| = m \leqslant n$. Let $g \in G$. Then $gH = \{gh: h \in H\}$ and hence $|gH| = |H| = m$. Thus every left coset of H contains exactly m elements.

Let $\Sigma = \{X_1, X_2, \ldots, X_r\}$ be the set of distinct left cosets of H in G. Then, from corollary 7.2.2, Σ partitions G and hence every element of G belongs to exactly one of the sets $X_1, X_2, \ldots, X_r$. Thus

$$|G| = |X_1| + |X_2| + \ldots + |X_r|$$
$$= rm.$$

Thus m is a divisor of the order of G. ∎

Index of a subgroup

The number of distinct left cosets of a subgroup H in a group G is called the *index* of H in G.

In the proof of Lagrange's theorem, the index of H in G was denoted by the integer r. Since we established $|G| = mr$ we have the following relationship when G is a finite group:

$$\text{index of } H \text{ in } G = \frac{|G|}{|H|} .$$

From Lagrange's theorem we can deduce the following two corollaries. They give an indication of the power of the theorem. In fact, corollary 7.2.5 completely determines the structure of any group whose order is a prime number. We look at this again in section 7.4.

COROLLARY 7.2.4 Let G be a finite group of order n. Then the order of every element of G is a divisor of n.

Proof Suppose $g \in G$ has order m. Then $G = \langle g \rangle$ also has order m (result 6.4.1). Hence by Lagrange's theorem, m is a divisor of n. ∎

COROLLARY 7.2.5 Every group of prime order p is cyclic.

Proof Since $p > 1$, G contains a non-identity element, g, say. Let $o(g) = m > 1$. Then by corollary 7.2.4, $m \mid p$ and hence, since p is prime, $m = p$. Thus G is cyclic, generated by any non-identity element. ∎

PROBLEMS 7.2a

1 Find the set of (a) distinct left cosets (b) distinct right cosets of the subgroup $K = \{\iota, (1 \ \ 2)\}$ in S_3. Comment on whether

$$\gamma K = K\gamma, \ \forall \ \gamma \in S_3.$$

2 The alternating group $A_4 = \{\iota, (1 \ \ 2)(3 \ \ 4), \ (1 \ \ 3)(2 \ \ 4), (1 \ \ 4)(2 \ \ 3), (1 \ \ 2 \ \ 3), (1 \ \ 3 \ \ 2), (1 \ \ 2 \ \ 4), (1 \ \ 4 \ \ 2), (1 \ \ 3 \ \ 4), (1 \ \ 4 \ \ 3), (2 \ \ 3 \ \ 4), (2 \ \ 4 \ \ 3)\}$.
   Repeat problem 1 for the subgroup $V = \{\iota, (1 \ \ 2)(3 \ \ 4), (1 \ \ 3)(2 \ \ 4), (1 \ \ 4)(2 \ \ 3)\}$ in A_4.

3 A subgroup N of a group G is called a *normal* subgroup if $xN = Nx$ for every element $x \in G$. Decide which of the following subgroups are *normal* in the given group:

   (a) $H_| = \{\bar{0}, \bar{3}\}$ in $(\mathbb{Z}_6, \oplus)$;
   (b) $H = \{\iota, (1 \ \ 2 \ \ 3), (1 \ \ 3 \ \ 2)\}$ in S_3;
   (c) $K = \{\iota, (1 \ \ 2)\}$ in S_3;
   (d) $T = \{\iota, (1 \ \ 2), (3 \ \ 4), (1 \ \ 2)(3 \ \ 4)\}$ in S_4.

4 Prove that every subgroup of an abelian group is normal.

5 Let G be a group. The subset C of elements of G which commute with *every* element of G is called the *centre* of G.
Thus $C = \{x \in G: xg = gx, \; \forall \, g \in G\}$. Prove that

(a) C is non-empty;   (b) C is a subgroup of G;
(c) C is normal in G.

What is the centre of S_3?

6 Let $H = \{e, h\}$ be a subgroup of the group G. Prove that H is normal in G if and only if $h \in C$, the centre of G.

7 Find the index of each of the following subgroups in $(\mathbb{Z}, +)$:

(a) $2\mathbb{Z} = \{2r: r \in \mathbb{Z}\}$;   (b) $5\mathbb{Z} = \{5r: r \in \mathbb{Z}\}$.

8 Let H be a proper subgroup of a group G of order 12. Give the possibilities for the order of H.

(a) Find a subgroup of $(\mathbb{Z}_{12}, \oplus)$ of each of these orders.
(b) Find, where possible, a subgroup of A_4 of each of these orders.

Is it true that a group necessarily contains a subgroup of each of the orders that is possible by Lagrange's theorem?

9 A group G has subgroups H, K. Find the possibilities for the order of the subgroup $H \cap K$ when

(a) $|H| = 16$, $|K| = 20$;   (b) $|H| = 3$, $|K| = 7$;
(c) $|H| = |K| = 7$.

In cases (b) and (c) find, also, the number of elements in the set $H \cup K$.

10 Let G be a non-cyclic group of order p^2 where p is a prime. Prove that every non-identity element of G has order p.

11 Let K, L be subsets of a group G. The subset KL of G is defined by

$$KL = \{kl: k \in K, l \in L\}.$$

Find KL when G is the group S_3, $K = \{\iota, (1 \; 2)\}$ and

(a) $L = \{\iota, (1 \; 3)\}$;   (b) $L = \{\iota, (1 \; 2)\}$;
(c) $L = \{(1 \; 3), (2 \; 3), (1 \; 2 \; 3)\}$.

Is it true that if K and L are both groups, then KL is a group? Can anything be said about the maximum and minimum values of $|KL|$ in terms of $|K|$ and $|L|$?

PROBLEMS 7.2b

1 Find the set of (a) distinct left cosets and (b) distinct right cosets of the subgroup $H = \{\iota, (1 \ \ 2 \ \ 3), (1 \ \ 3 \ \ 2)\}$ in the alternating group A_4 (see problem 7.2a, 2). Comment on whether $\gamma H = H\gamma$, $\forall \ \gamma \in A_4$.

2 Decide which, if any, of the following subgroups are *normal* in A_4 (see problem 7.2a, 3).
   (a) $V = \{\iota, (1 \ \ 2)(3 \ \ 4), (1 \ \ 3)(2 \ \ 4), (1 \ \ 4)(2 \ \ 3)\}$;
   (b) $H = \{\iota, (1 \ \ 2 \ \ 3), (1 \ \ 3 \ \ 2)\}$;
   (c) $L = \{\iota, (1 \ \ 2)(3 \ \ 4)\}$.

3 Let H be a subgroup of a group G. Prove that $hH = H$, $\forall \ h \in H$.

4 Let G be a group of order $2m$ containing a subgroup H of order m. Describe the elements in each of the two distinct left cosets of H in G. Hence prove that H is normal in G.

5 Let G be a group of order n. Prove that $g^n = e$ for every element $g \in G$.

6 Let G be a group of order 8. Prove that either G is cyclic or $g^4 = e$, for every element $g \in G$, where e is the identity element of G.

7 Suppose that H is a subgroup of K and K is a subgroup of G. What are the possibilities for the order of K if $|H| = 4$ and $|G| = 24$?

8 Subsets J and H of the set $\mathbb{C}$ of complex numbers are defined by $J = \{a + ib: a, b \in \mathbb{Z}\}$ and $H = \{2a + 3ib: a, b \in \mathbb{Z}\}$. Given that $(J, +)$ and $(H, +)$ are subgroups of $(\mathbb{C}, +)$, show that the index of $(H, +)$ in $(J, +)$ is 6.

9 A group G has distinct subgroups H, K. Find the number of elements in the subgroup $H \cap K$ and in the set $H \cup K$ when

   (a) $|H| = p$, $|K| = q$, where p, q are distinct prime numbers;
   (b) $|H| = |K| = p$, where p is a prime.

   Suppose $|G| = 10$. Prove that G must contain an element of order 2 and an element of order 5.

10 Let H be a subgroup of a group G. The subset $a^{-1}Ha$, where $a \in G$, is defined by

$$a^{-1}Ha = \{a^{-1}ha: h \in H\}.$$

   (a) Prove that $a^{-1}Ha$ is a subgroup of G.
   (b) Prove that H is normal if and only if $a^{-1}Ha = H$, for all $a \in G$.
   (c) Hence prove that if H is a normal subgroup of G, then given any $h \in H$ and any $a \in G$, there exists $h_1, h_2 \in H$ such that

$$ah = h_1 a \quad \text{and} \quad ha = ah_2.$$

11 Let N be a normal subgroup of a group G. Prove that

$$(aN)(bN) = abN, \quad \text{for every } a, b \in G,$$

where $(aN)(bN) = \{xy: x \in aN, y \in bN\}$.

7.3 DIRECT PRODUCTS

Let G, H be groups and let S be the cartesian product set $G \times H$.

That is, $\qquad S = \{(g, h): g \in G, h \in H\}$.

We define a binary operation on S by the rule

$$(g_1, h_1) * (g_2, h_2) = (g_1 g_2, h_1 h_2),$$

where we are combining the first elements in the ordered pair by the *operation in G* and the second elements by the *operation in H*.

Examples 7.3.1

1. Let $S = (\mathbb{R}, +) \times (\mathbb{R}, +)$. Then $S = \{(a, b): a, b \in \mathbb{R}\}$ and $(a_1, b_1) * (a_2, b_2) = (a_1 + a_2, b_1 + b_2)$. This is just the law of vector addition in $\mathbb{R}^2$.

2. Let $G = \{e, a\}$, $H = \{f, b, b^2\}$ be cyclic groups of orders 2 and 3 with identity elements e, f respectively. Then the Cayley table for $(S, *)$ is as follows:

$*$	(e, f)	(e, b)	(e, b^2)	(a, f)	(a, b)	(a, b^2)
(e, f)	(e, f)	(e, b)	(e, b^2)	(a, f)	(a, b)	(a, b^2)
(e, b)	(e, b)	(e, b^2)	(e, f)	(a, b)	(a, b^2)	(a, f)
(e, b^2)	(e, b^2)	(e, f)	(e, b)	(a, b^2)	(a, f)	(a, b)
(a, f)	(a, f)	(a, b)	(a, b^2)	(e, f)	(e, b)	(e, b^2)
(a, b)	(a, b)	(a, b^2)	(a, f)	(e, b)	(e, b^2)	(e, f)
(a, b^2)	(a, b^2)	(a, f)	(a, b)	(e, b^2)	(e, f)	(e, b)

It is easy to check that $(S, *)$ satisfies the group axioms.

> **THEOREM 7.3.1** Let G, H be groups. Then the cartesian product set $S = G \times H$ is a group with respect to the operation $*$ defined by
>
> $$(g_1, h_1) * (g_2, h_2) = (g_1 g_2, h_1 h_2).$$

Proof Axiom 1   Let (g_1, h_1), $(g_2, h_2) \in S$. Then $g_1 g_2 \in G$ and $h_1 h_2 \in H$ since G, H are groups. Hence $(g_1 g_2, h_1 h_2) \in S$.

Axiom 2   Using the associative law in G and H gives

$$
\begin{aligned}
(g_1, h_1) * (g_2 g_3, h_2 h_3) &= (g_1(g_2 g_3), h_1(h_2 h_3)) \\
&= ((g_1 g_2)g_3, (h_1 h_2)h_3) \\
&= (g_1 g_2, h_1 h_2) * (g_3, h_3).
\end{aligned}
$$

Axiom 3   Let e, f be the identity elements of G, H respectively. Then since $(eg, fh) = (g, h) = (ge, hf)$, for all $g \in G$, $h \in H$, (e, f) is the identity of $(S, *)$.

Axiom 4   Let $(g, h) \in S$. Then $g \in G$, $h \in H \Rightarrow g^{-1} \in G$, $h^{-1} \in H \Rightarrow (g^{-1}, h^{-1}) \in S$.
Since $(g^{-1}g, h^{-1}h) = (e, f) = (g\, g^{-1}, h\, h^{-1})$, (g^{-1}, h^{-1}) is the inverse of (g, h).
Hence $(S, *)$ is a group called the *direct product* of G and H. ■

We can extend this definition to any finite number of groups.
Let $G_1, G_2, \ldots, G_n$ be groups and let $S = G_1 \times G_2 \times , \ldots , \times G_n$.
Define the binary operation $*$ on S by

$$(x_1, x_2, \ldots, x_n) * (y_1, y_2, \ldots, y_n) = (x_1 y_1, x_2 y_2, \ldots, x n y_n).$$

The proof given in theorem 7.3.1 can easily be adapted to show that $(S, *)$ is a group.

As we have done previously, we will usually drop the $*$ and refer simply to the *direct product* $G_1 \times G_2 \times , \ldots , \times G_n$.

PROBLEMS 7.3a

1 Let G, H be the cyclic groups defined in example 7.3.1, 2. Find the order of each element of $G \times H$. Deduce that $G \times H$ is cyclic and define an isomorphic mapping of $G \times H$ onto $(\mathbb{Z}_6, \oplus)$.

2 Let $G = \{e, a\}$ and $H = \{e, b\}$ both be subgroups of a group with identity element e. Show that $G \times H \cong V$, the Klein 4-group.

3 Let G, H be abelian groups. Prove that $G \times H$ is abelian. State the *converse* result. Decide whether the converse is true, giving a proof or counter-example, as appropriate.

4 List the elements of $S = (\mathbb{Z}_2, \oplus) \times (\mathbb{Z}_4, \oplus)$ and find the order of each. Is S isomorphic to D_4, the octic group? Justify your answer.

5 Find the least integer m such that g^m is the identity for every $g \in G$, and an element of maximum order in G where G is

(a) $(\mathbb{Z}_2, \oplus) \times (\mathbb{Z}_5, \oplus)$; (b) $(\mathbb{Z}_2, \oplus) \times (\mathbb{Z}_6, \oplus)$; (c) $(\mathbb{Z}_4, \oplus) \times (\mathbb{Z}_6, \oplus)$.

6 Let $\mathbb{R}^+ = \{x \in \mathbb{R} : x > 0\}$ and $U = \{1, -1\}$. Prove that

$$(\mathbb{R}^*, \times) \cong (U, \times) \times (\mathbb{R}^+, \times).$$

7 Let $\mathbb{Z}_{15}^*$ denote the set of residues (mod 15) that are *coprime* with 15. Show that every element of the group $(\mathbb{Z}_{15}^*, \otimes)$ can be expressed uniquely in the form $a^r b^s$, $r = 0, 1$; $s = 0, 1, 2, 3$; where $a = \overline{11}$ and $b = \overline{2}$. Let $A = \langle \overline{11} \rangle$ and $B = \langle \overline{2} \rangle$. Construct an isomorphic mapping of $(\mathbb{Z}_{15}^*, \otimes)$ onto $A \times B$. Hence prove that $(\mathbb{Z}_{15}^*, \otimes) \cong (\mathbb{Z}_2, \oplus) \times (\mathbb{Z}_4, \oplus)$.

PROBLEMS 7.3b

1 Show that the group $(\mathbb{Z}_3, \oplus) \times (\mathbb{Z}_4, \oplus)$ is cyclic and find all the generators.

2 List the elements of $S = (\mathbb{Z}_2, \oplus) \times (\mathbb{Z}_2, \oplus) \times (\mathbb{Z}_2, \oplus)$ and give the order of each. The groups S, $(\mathbb{Z}_8, \oplus)$ and $(\mathbb{Z}_4, \oplus) \times (\mathbb{Z}_2, \oplus)$ are all abelian groups of order 8. Is any pair isomorphic? Justify your answer.

3 Prove that $G = (\mathbb{Z}_m, \oplus) \times (\mathbb{Z}_n, \oplus)$ is cyclic if $\gcd(m, n) = 1$. Suppose that $\gcd(m, n) = d$. Prove that the order of every element of G divides mn/d and deduce that G is not cyclic if $d > 1$.

4 Let H_1, H_2 be groups isomorphic to G_1, G_2 respectively. Prove that $H_1 \times H_2 \cong G_1 \times G_2$.

5 Prove that $(\mathbb{C}, +) \cong (\mathbb{R}, +) \times (\mathbb{R}, +)$.

6 Find all the elements of $(\mathbb{R}^*, \times) \times (\mathbb{R}^*, \times)$ which have *finite* order. Hence prove that $(\mathbb{C}^*, \times)$ is not isomorphic to $(\mathbb{R}^*, \times) \times (\mathbb{R}^*, \times)$.

7 Let $\mathbb{Z}_{21}^*$ denote the set of residues (mod 21) which are coprime with 21. Prove that $(\mathbb{Z}_{21}^*, \otimes) \cong (\mathbb{Z}_2, \oplus) \times (\mathbb{Z}_6, \oplus)$.

7.4 ABSTRACT GROUPS

You might wonder whether we know how to construct all the possible types of group of a given order n. The answer is that we do not; although we know how the abelian groups of order n should be constructed, we do not know in all cases how to construct the non-abelian groups, although a great stride has recently been made with the publication of a classification of all the *simple* groups (groups which have no normal subgroups). (For a readable account of the history, methods and importance of this classification see D. Gorenstein, 'The enormous theorem', *Scientific American*, Dec. 1985, 92–103.) We conclude our brief introduction to the theory of groups by looking at the problem of constructing groups of order n in some particularly simple cases. First, some general results.

THEOREM 7.4.1   Let G be a group of prime order p. Then
(a) G is cyclic;
(b) $G \cong (\mathbb{Z}_p, \oplus)$.
(c) G contains an identity element and $p-1$ elements of order p;
(d) every non-identity element is a generator;
(e) G is abelian.

Proof (a) This was proved as corollary 7.2.5 to Lagrange's theorem.
(b) This follows from theorem 7.1.7.
(c) and (d)   By corollary 7.2.4 to Lagrange's theorem, all non-identity elements have order p and hence are generators.
(e) This follows from (a) and result 6.4.4. ∎

Generating set

Suppose G is a group. If G is non-cyclic, no single element generates G; however, we can always choose a subset $X \subseteq G$ such that every element of G can be expressed as the product of powers (positive, negative or zero) of the elements of X. It is clear that X exists because trivially we could choose $X = G$. Such a set X is called a *generating set* for G. Now X may contan redundant elements in the sense that one or more of its elements can be expressed as a product of powers of some of the remaining elements of X. If we weed out all such redundant elements the resulting set is called a *minimal generating set* for G.

Notation

Let $X = \{a_1, a_2, \ldots, a_m\}$ be a generating set for a group G. Then we write $G = \langle X \rangle$ or $G = \langle a_1, a_2, \ldots, a_m \rangle$.

Examples 7.4.1

1. The minimal generating set for a *cyclic* group consists of a single element.

2. Let $V = \{e, a, b, c\}$ be a Klein 4-group. Then V is non-cyclic so that any generating set must contain more than one element. Clearly $S = \{a, b, c\}$ is a generating set, since $e = a^0$ (or a^2). However it is not *minimal*, since $a = bc$, for example. Hence $X_1 = \{b, c\}$ is a minimal generating set for V. Other minimal generating sets for V are $X_2 = \{a, c\}$ and $X_3 = \{a, b\}$.

3. Let $\alpha = \begin{pmatrix} 1 & 2 & 3 \\ 2 & 3 & 1 \end{pmatrix}$ and $\beta = \begin{pmatrix} 1 & 2 & 3 \\ 1 & 3 & 2 \end{pmatrix}$. Then the six elements of S_3 can be expressed as $\{\alpha^0, \alpha, \alpha^2, \beta, \alpha\beta, \alpha^2\beta\}$. Hence $S_3 = \langle \alpha, \beta \rangle$. Since S_3 is non-cyclic, $\langle \alpha, \beta \rangle$ must be a *minimal* generating set for S_3. There are other minimal generating sets for S_3 (see problem 7.4a, 2).

Abelian groups

Suppose that $G = \langle a, b, c, \ldots \rangle$ is an abelian group and $g \in G$. Then since every pair of generators commute, we can express g in the form $g = a^r b^s c^t, \ldots$, where $r, s, t, \ldots \in \mathbb{Z}$, by collecting together the powers of each generator, just as in elementary algebra where the letters are standing for real numbers.

RESULT 7.4.2 Let G by a group in which every non-identity element has order 2. Then G is abelian.

Proof Every element $x \in G$ satisfies the equation $x^2 = e$ and hence $x = x^{-1}$, $\forall x \in G$. Let a, b be any two elements of G. Then $ab \in G$ and hence $ab = (ab)^{-1} = b^{-1}a^{-1} = ba$. Thus G is abelian. ∎

Notation

For simplicity, we shall denote by C_n the cyclic group of order n, written multiplicatively. Thus $C_n = \langle a \rangle = \{e, a, \ldots, a^{n-1}\}$, where $a^n = e$. As we showed in theorem 7.1.7, $C_n \cong (\mathbb{Z}_n, \oplus)$.

THEOREM 7.4.3 Let G be a non-trivial group in which every non-identity element has order 2.
Then $|G| = 2^m$ and $G \cong C_2 \times C_2 \times, \ldots, \times C_2$.

Proof Let $S = \{a_1, a_2, \ldots, a_m\}$ be a minimal generating set for G and let $g \in G$. Then since G is abelian by result 7.4.2, we can write $g = a_1{}^{r_1} a_2{}^{r_2} \ldots a_m{}^{r_m}$, where $0 \leq r_i < o(a_i)$. Since e would be redundant in a generating set for any non-trivial group, we can assume $o(a_i) = 2$, $i = 1, 2, \ldots, m$. Then we can take $r_i \in \{0, 1\}$, $i = 1, 2, \ldots, m$.

We next show that each element of G has a unique expression in this form. Suppose for some $g \in G$ we have $g = a_1{}^{r_1} a_2{}^{r_2} \ldots a_m{}^{r_m} = a_1{}^{s_1} a_2{}^{s_2} \ldots a_m{}^{s_m}$, where $r_i, s_i \in \{0, 1\}$, $i = 1, 2, \ldots, m$. Then premultiplying both sides of the second equality successively by $a_1{}^{-s_1}, a_2{}^{-s_2}, \ldots, a_m{}^{-s_m}$ and using the fact that the generators commute, gives $a_1{}^{t_1} a_2{}^{t_2} \ldots a_m{}^{t_m} = e$, where $t_i = r_i - s_i$ and hence $|t_i| \in \{0, 1\}$, $i = 1, 2, \ldots, m$. If t_i were non-zero for *more than one* value of i we could express one generator as a product of powers of some of the other generators, and if t_i were non-zero for *just one* value of i then e would be a generator, both contrary to our assumption that S is a *minimal* generating set. Hence $t_i = 0$, $i = 1, 2, \ldots, m$, and the expression for g in the form $a_1{}^{r_1} a_2{}^{r_2} \ldots a_m{}^{r_m}$ is unique. Hence since there are exactly two choices for each of $r_1, r_2, \ldots, r_m$, $|G| = 2^m$. Also, since $o(a_i) = 2$, $\langle a_i \rangle \cong C_2$, $i = 1, 2, \ldots, m$, and so a typical element in the direct product group $C_2 \times C_2 \times \ldots \times C_2$ can be written as $(a_1{}^{r_1}, a_2{}^{r_2}, \ldots, a_m{}^{r_m})$, where $r_i \in \{0, 1\}$, and the mapping $\phi: G \to C_2 \times C_2 \times \ldots \times C_2$ defined by $\phi: a_1{}^{r_1} a_2{}^{r_2} \ldots a_m{}^{r_m} \mapsto (a_1{}^{r_1}, a_2{}^{r_2}, \ldots, a_m{}^{r_m})$ is a bijection.

Finally, we show that ϕ is an isomorphism. Let $g = a_1{}^{r_1} a_2{}^{r_2} \ldots a_m{}^{r_m}$, $h = a_1{}^{q_1} a_2{}^{q_2} \ldots a_m{}^{q_m}$. Then, since the generators of G commute, $gh = a_1{}^{r_1 + q_1} a_2{}^{r_2 + q_2} \ldots a_m{}^{r_m + q_m}$ and hence $\phi(gh) = (a_1{}^{r_1 + q_1}, a_2{}^{r_2 + q_2} \ldots, a_m{}^{r_m + q_m}) = \phi(g)\phi(h)$. Thus $G \cong C_2 \times C_2 \times \ldots \times C_2$, completing the proof of the theorem. ∎

COROLLARY 7.4.4 Let G be a group of order $2p$, where p is an odd prime, then G contains an element of order p.

Proof Since $|G| \neq 2^m$, G contains a non-identity element g of order greater than 2. But since the order of g divides $|G| = 2p$, either $o(g) = p$ or $o(g) = 2p$. In the latter case, it is easy to see that g^2 is an element of order p. Thus in either case G contains an element of order p. ∎

Finally, we indicate how to construct all the groups of order n, for $n < 12$. From theorem 7.4.1, we know that there is only one type of group of each of the orders 2, 3, 5, 7, 11 and we know its structure. In theorems 7.4.6 and 7.4.7 we will prove that there are just two types of group of each of the orders 4 and 6. We leave the reader to explore the cases $n = 8, 9, 10$ in the problems 7.4a and 7.4b. The case when $n = 9$ is simplified by the following result:

THEOREM 7.4.5 Let G be a group of order p^2, where p is a prime. Then G is abelian.

The proof of theorem 7.4.5, although not difficult, is beyond the scope of this book. See Herstein's *Topics in Algebra* or any of the other books on groups listed in the Bibliography.

THEOREM 7.4.6 A group G of order 4 is either cyclic or isomorphic to V, the Klein 4-group.

Proof By corollary 7.2.4 to Lagrange's theorem, the order of every non-identity element of G is either 2 or 4. If G contains an element of order 4 then G is cyclic. Otherwise, every non-identity element of G has order 2 and hence by theorem 7.4.3, $G \cong C_2 \times C_2 \cong V$ (see problem 7.3a, 2). ∎

THEOREM 7.4.7 A group G of order 6 is either cyclic or isomorphic to S_3.

Proof By corollary 7.2.4 to Lagrange's theorem, the order of every non-identity element of G is 2, 3 or 6. If G contains an element of order 6, then G is cyclic. Suppose every non-identity element of G has order 2 or 3. Then by corollary 7.4.4, G contains an element b, say, of order 3. So G contains the subgroup $K = \langle b \rangle = \{e, b, b^2\}$. Since $|G| = 6$, K has two distinct left cosets in g. One of these cosets is $eK = K$. Suppose the other is aK, where $a \in G - K$. Then $G = eK \cup aK = \{e, b, b^2, a, ab, ab^2\}$. But products such as $ba, a^2, b^2a, bab, \ldots$ all belong to G, so that to complete the Cayley table for G we need to be able to identity these among the list of elements of G. It is not difficult to see that if we can identify just a^2 and ba, then we can complete the table for G. Since $a \notin K$, then $a^2 \notin aK$ and hence a^2 is one of the elements e, b or b^2. If $a^2 \neq e$, then a has order 3, by assumption. But $a^2 = b \Rightarrow a^3 = ab \neq e$

and $a^2 = b^2 \Rightarrow a^3 = ab^2 \neq e$. This contradiction shows that $a^2 = e$ and a has order 2.

Now consider ba. It is easy to see that ba is not one of the elements e, b, b^2 or a (see problems 7.4a, 3). We are left with the two possibilities $ba = ab$ or $ba = ab^2$. Now $ba = ab \Rightarrow$ every pair of elements in G commutes and G is abelian. Hence $(ab)^2 = a^2b^2 = b^2 \neq e$ and $(ab)^3 = abb^2 = a \neq e$, so that ab is not an element of order 2 or 3, contradicting our assumption. Hence the only possibility is $ba = ab^2$. It is easy to check that in this case ab and ab^2 both have order 2. Thus there is just one possibility for a non-cyclic group of order 6 and this is a non-abelian group $G = \langle a, b \rangle$ where $a^2 = e = b^3$ and $ba = a^2b$. This information is sufficient for us to complete a Cayley table which is easily seen to be isomorphic to the table for S_3. One possible isomorphism $\phi: G \to S_3$ is given by $\phi(b) = \begin{pmatrix} 1 & 2 & 3 \\ 2 & 3 & 1 \end{pmatrix}$ and $\phi(a) = \begin{pmatrix} 1 & 2 & 3 \\ 1 & 3 & 2 \end{pmatrix}$. ∎

PROBLEMS 7.4a

1 How many different minimal generating sets are there for $(\mathbb{Z}_{17}, \oplus)$?

2 Decide which of the following subsets are generating sets for S_3:

(a) any two of the elements of order 2;
(b) the two elements of order 3;
(c) any one of the elements of order 2 and any one of the elements of order 3;
(d) any three elements of S_3.

Find the number of minimal generating sets for S_3.

3 Suppose that $G = \{e, b, b^2, a, ab, ab^2\}$ is a group with identity element e and the elements a, b are of order 2 and 3 respectively. Prove that

(a) ba is not one of the elements e, b, b^2 or a;
(b) if $ba = ab$, then G is cyclic;
(c) if $ba = ab^2$, then ab and ab^2 both have order 2.

4 Suppose that G is a group of order 8 which is not isomorphic to either C_8 or to $C_2 \times C_2 \times C_2$. Explain why G must contain an element b of order 4. Show that the elements of G can be written as $\{e, b, b^2, b^3, a, ab, ab^2, ab^3\}$, where $a \notin \langle b \rangle$, and prove that $a^2 = e$ or $a^2 = b^2$. Prove that in either case, the only possibilities for ba are $ba = ab$ or $ba = ab^3$.

(a) Assume $a^2 = e$ and $ba = ab$. Construct an isomorphism from G onto $C_2 \times C_4$.

(b) Assume $a^2 = b^2$ and $ba = ab$. Let $c = ab$ and show $o(c) = 2$ and that the elements of G can be rewritten as
$\{e, b, b^2, b^3, c, cb, cb^2, cb^3\}$. Deduce that $G \cong C_2 \times C_4$.

(c) Assume $a^2 = e$ and $ba = ab^3$. Construct the Cayley table for G and hence show that $G \cong D_4$, the octic group (see problem 6.7a, 5).

(d) Verify that the set $M = \{I, B, B^2, B^3, A, AB, AB^2, AB^3\}$ of matrices over $\mathbb{C}$ form a multiplicative group, where

$$A = \begin{pmatrix} 0 & -1 \\ 1 & 0 \end{pmatrix} \text{ and } B = \begin{pmatrix} i & 0 \\ 0 & -1 \end{pmatrix}.$$

Verify also that $B^4 = I$, $A^2 = B^2 \neq I$, and $BA = AB^3$. This group is called the *quaternion group*.

Deduce that there are just five types of group of order 8.

PROBLEMS 7.4b

1 Let $G = \langle a_1, a_2, \ldots, a_m \rangle$, where $a_i a_j = a_j a_i$ for all $1 \leq i < j \leq m$. Prove that G is abelian.

2 Let G be a group in which every non-identity element has order 2 and suppose $\{a_1, a_2, \ldots, a_m\}$ is a minimal generating set for G. Define an isomorphic mapping from G onto the group $(\mathbb{Z}_2, \oplus) \times (\mathbb{Z}_2, \oplus) \times \ldots \times (\mathbb{Z}_2, \oplus)$, where there are m terms in the product.

3 Complete the definition of the isomorphism $\phi: G \to S_3$ given at the end of the proof of theorem 7.4.7 by calculating the images under ϕ of the remaining elements of G.

4 Let G be a group of order 9. Prove that if $a, b \in G$ are both elements of order 3, then either $\langle a \rangle = \langle b \rangle$ or $\langle a \rangle \cap \langle b \rangle = \{e\}$. Suppose G contains no element of order 9. Explain why every non-identity element has order 3 and deduce that given $K = \langle b \rangle$, where $b \in G$, $b \neq e$, then we can find $a \in G - K$ such that $\{eK, aK, a^2K\}$ is a complete set of left cosets of K in G. Hence show that the elements of G can be written in the form $G = \{a^r b^s: 0 \leq r \leq 2, 0 \leq s \leq 2\}$, where $a^3 = e = b^3$. Assuming theorem 7.4.5, construct an isomorphism from G onto $C_3 \times C_3$. Deduce that there are just two types of group of order 9.

5 Let G be a group of order 10. Explain why G contains an element b of order 5. Let $K = \langle b \rangle$ and prove that if $\{eK, aK\}$ is a complete set of left cosets of K in G, then $a^2 = e$. Prove also that if $ba = ab$, then G is cyclic. Suppose G contains no element of order 10. Prove that $ba = ab^4$ is the only possibility and that in this case G is isomorphic to the dihedral group D_5 (see problem 6.7b, 2). Deduce that there are just two types of group of order 10.

SOLUTIONS TO PROBLEMS IN THE (a) SETS

Problems 7.1a

1 6;

	$\bar{1}$	$\bar{3}$	$\bar{5}$	$\bar{7}$
ϕ_1	ι	A	B	C
ϕ_2	ι	A	C	B
ϕ_3	ι	B	A	C
ϕ_3	ι	B	C	A
ϕ_5	ι	C	A	B
ϕ_6	ι	C	B	A

2 (a)

x	$\bar{1}$	$\bar{3}$	$\bar{7}$	$\bar{9}$
$\phi_1(x)$	$\bar{0}$	$\bar{1}$	$\bar{3}$	$\bar{2}$
or $\phi_2(x)$	$\bar{0}$	$\bar{3}$	$\bar{1}$	$\bar{2}$

(b)

x	1	-1	i	$-i$
$\phi_1(x)$	$\bar{0}$	$\bar{2}$	$\bar{1}$	$\bar{3}$
or $\phi_2(x)$	$\bar{0}$	$\bar{2}$	$\bar{3}$	$\bar{1}$

(c)

x	$\bar{1}$	$\bar{5}$	$\bar{7}$	$\overline{11}$
$\phi(x)$	e	a	b	c

3 From results 7.1.1 and 7.1.3, the result holds when $m \geqslant 0$.
Now suppose $m < 0$. Put $m = -n$, where $n > 0$. Then
$x^m = x^{-n} = (x^{-1})^n$, by definition. But $\phi(x^{-1}) = y^{-1}$ by result 7.1.2
and hence $\phi(x^{-1})^n = (y^{-1})^n$ by result 7.1.3, giving $\phi(x^{-n}) = y^{-n}$.
So $= \phi(x^m) = y^m$ for all $x \in G$, $m \in \mathbb{Z}$.

4

x	e	a	a^2	a^3	a^4	a^5
$\phi_1(x)$	$\bar{0}$	$\bar{1}$	$\bar{2}$	$\bar{3}$	$\bar{4}$	$\bar{5}$
$\phi_2(x)$	$\bar{0}$	$\bar{5}$	$\bar{4}$	$\bar{3}$	$\bar{2}$	$\bar{1}$

No; because $\phi(a)$ must be an element of order 6 in $(\mathbb{Z}_6, \oplus)$. Once
$\phi(a)$ is decided, $\phi(a^n)$ is uniquely determined by result 7.1.3,
$n = 2, 3, 4, 5$.

5 Any automorphism of $(\mathbb{Z}_6, \oplus)$ must map $\bar{1}$ into an element of order
6. This gives two possibilities: $\phi(\bar{1}) = \bar{1}$ or $\phi(\bar{1}) = \bar{5}$.

(a) $(\mathbb{Z}_3, \oplus)$:

x	$\bar{0}$	$\bar{1}$	$\bar{2}$
$\phi_1(x)$	$\bar{0}$	$\bar{1}$	$\bar{2}$
$\phi_2(x)$	$\bar{0}$	$\bar{2}$	$\bar{1}$

(b) $(\mathbb{Z}_4, \oplus)$:

x	$\bar{0}$	$\bar{1}$	$\bar{2}$	$\bar{3}$
$\phi_1(x)$	$\bar{0}$	$\bar{1}$	$\bar{2}$	$\bar{3}$
$\phi_2(x)$	$\bar{0}$	$\bar{3}$	$\bar{2}$	$\bar{1}$

6

α	ι	(1 2 3)	(1 3 2)	(2 3)	(1 3)	(1 2)
$\phi_1(\alpha)$	ι	(1 2 3)	(1 3 2)	(1 3)	(1 2)	(2 3)
$\phi_2(\alpha)$	ι	(1 2 3)	(1 3 2)	(1 2)	(2 3)	(1 3)
$\phi_3(\alpha)$	ι	(1 3 2)	(1 2 3)	(1 2)	(1 3)	(2 3)

7 2; $(\mathbb{Z}, +)$ has just two generators, 1 and -1. Hence an automorphism of $(\mathbb{Z}, +)$ must map 1 into either 1 or -1. Hence the two automorphisms are $\phi_1(x) = x$, $x \in \mathbb{Z}$; $\phi_2(x) = -x$, $x \in \mathbb{Z}$.

8 No; let $x_1, x_2 \in \mathbb{R}^*$. Then $\phi(x_1 x_2) = 5x_1 x_2$. But $\phi(x_1)\phi(x_2) = (5x_1)(5x_2) = 25x_1 x_2$. Hence $\phi(x_1 x_2) \neq \phi(x_1)\phi(x_2)$.

9 There are many possible proofs. We give two examples of proofs by contradiction.

(a) Use the fact that (-1) has two square roots in $(\mathbb{C}^*, \times)$ but none in $(\mathbb{R}^*, \times)$. The method is the same as that used in example 7.1.2, 2 with $\mathbb{C}^*$ in place of $\mathbb{R}^+$, $\mathbb{R}^*$ in place of $\mathbb{Q}^+$ and (-1) in place of 2.

(b) Use the existence of n nth roots of unity in $\mathbb{C}^*$, taking $n > 2$. Suppose $\phi \colon (\mathbb{C}^*, \times) \to (\mathbb{R}^*, \times)$ is an isomorphism. Then by result 7.1.1, $\phi(1 + 0i) = 1$. Let $\phi(-1 + 0i) = x_1$, $\phi(0 + i) = x_2$, $\phi(0 - i) = x_3$. Then in $(\mathbb{C}^*, \times)$, $(-1)^4 = i^4 = (-i)^4 = 1$. Hence in $(\mathbb{R}^*, \times)$, $x_1^4 = x_2^4 = x_3^4 = 1$. Since ϕ is injective, the numbers $1, x_1, x_2, x_3$ are all distinct. But this is impossible, since 1 has only two 4th roots (1 and -1) in $(\mathbb{R}^*, \times)$.

10 (i) Show ϕ is bijective: $\phi \colon \mathbb{R} \to \mathbb{R}^+$ has an inverse mapping $\phi^{-1} \colon \mathbb{R}^+ \to \mathbb{R}$ defined by $\phi^{-1}(x) = \log_2 x$. Hence ϕ is bijective.

(ii) Show ϕ satisfies $\phi(x + y) = \phi(x)\phi(y)$, for all $x, y \in \mathbb{R}$. Now $\phi(x + y) = 2^{x+y} = 2^x 2^y = \phi(x)\phi(y)$.

Hence ϕ is an isomorphism.

11 In W, $\begin{bmatrix} x & -x \\ 0 & 0 \end{bmatrix} \begin{bmatrix} y & -y \\ 0 & 0 \end{bmatrix} = \begin{bmatrix} xy & -xy \\ 0 & 0 \end{bmatrix}$. This suggests we should

define $\phi \colon W \mapsto \mathbb{Q}^*$ by $\phi \colon \begin{bmatrix} x & -x \\ 0 & 0 \end{bmatrix} \mapsto x$, $\forall\, x \in \mathbb{Q}^*$.

Then ϕ is clearly bijective and satisfies

$$\phi\left(\begin{bmatrix} x & -x \\ 0 & 0 \end{bmatrix}\begin{bmatrix} y & -y \\ 0 & 0 \end{bmatrix}\right) = \phi\left(\begin{bmatrix} xy & -xy \\ 0 & 0 \end{bmatrix}\right) = xy$$

$$= \phi\left(\begin{bmatrix} x & -x \\ 0 & 0 \end{bmatrix}\right) \phi\left(\begin{bmatrix} y & -y \\ 0 & 0 \end{bmatrix}\right).$$

12 True; let G be non-abelian and H an abelian group. Suppose $\phi: G \to H$ is an isomorphism. Then since G is non-abelian we can find elements x, y such that $xy \neq yx$. Let $\phi(x) = a$, $\phi(y) = b$. Then since ϕ is an isomorphism, $\phi(xy) = \phi(x)\phi(y) = ab$ and $\phi(yx) = \phi(y)\phi(x) = ba$. Since ϕ is injective, $xy \neq yx \Rightarrow ab \neq ba$. But this contradicts the hypothesis that H is abelian.

13 (a) The mapping θ^{-1} is invertible (since θ is the inverse mapping of θ^{-1}) and hence θ^{-1} is bijective. Let x, $y \in G_2$. We must show that $\theta^{-1}(xy) = \theta^{-1}(x)\theta^{-1}(y)$. Let $\theta^{-1}(x) = a$ and $\theta^{-1}(y) = b$. Then $\theta(a) = x$ and $\theta(b) = y$. Since θ is an isomorphism, $\theta(ab) = \theta(a)\theta(b) = xy$. Thus $\theta^{-1}(xy) = ab = \theta^{-1}(x)\theta^{-1}(y)$, as required.

(b) Since θ, ϕ are bijective, the composition mapping $\phi \circ \theta$ is bijective (theorem 3.3.1). Let x, $y \in G$ and suppose $\theta(x) = x'$, $\theta(y) = y'$. Then $\theta(xy) = \theta(x)\theta(y) = x'y'$, since θ is an isomorphism. Hence $(\phi \circ \theta)(xy) = \phi(x'y') = \phi(x')\phi(y')$ since ϕ is an isomorphism. But $\phi(x') = \phi(\theta(x)) = (\phi \circ \theta)(x)$. Hence $(\phi \circ \theta)(xy) = (\phi \circ \theta)(x)(\phi \circ \theta)(y)$, showing that $\phi \circ \theta$ is an isomorphism.

(c) The identity mapping $\iota: G \to G$ is clearly an isomorphism of any group.

(a), (b) and (c) show that isomorphism is symmetric, transitive and reflexive respectively. Hence isomorphism is an equivalence relation.

14 We must show that

(a) $\forall\ x_1, x_2 \in \phi(H),\ x_1x_2 \in \phi(H)$; (b) $\forall\ x \in \phi(H),\ x^{-1} \in \phi(H)$.

(a) Let $x_1, x_2 \in \phi(H)$. Then we can find $h_1, h_2, \in H$ such that $\phi(h_1) = x_1$ and $\phi(x_2) = h_2$, by definition of $\phi(H)$. Then $h_1h_2 \in H$, since H is a subgroup. Hence $\phi(h_1h_2) \in \phi(H)$. But $\phi(h_1h_2) = \phi(h_1)\ \phi(h_2) = x_1x_2$. Thus $x_1x_2 \in \phi(H)$.

(b) Let $x \in \phi(H)$. Then we can find $h \in H$ such that $\phi(h) = x$. But $h^{-1} \in H$, since H is a subgroup. Hence $\phi(h^{-1}) \in \phi(H)$. But $\phi(h^{-1}) = x^{-1}$, by result 7.1.2. Hence $x^{-1} \in \phi(H)$. Thus $\phi(H)$ is a subgroup of G'.

Subgroups H and $\phi(H)$ have the same order. Since ϕ is bijective, it establishes a one–one correspondence between the elements of H and the elements of $\phi(H)$. Hence H is an infinite subgroup $\Leftrightarrow \phi(H)$ is an infinite subgroup and if H is finite, then $|H| = |\phi(H)|$.

Problems 7.2a

1 (a) $\{\iota, (1\ \ 2)\}$, $\{(1\ \ 3), (1\ \ 2\ \ 3)\}$, $\{(2\ \ 3), (1\ \ 3\ \ 2)\}$.

(b) $\{\iota, (1\ \ 2)\}$, $\{(1\ \ 3), (1\ \ 3\ \ 2)\}$, $\{(2\ \ 3), (1\ \ 2\ \ 3)\}$.

No; $\gamma K \neq K\gamma$ when $\gamma = (1\ \ 3), (2\ \ 3), (1\ \ 2\ \ 3)$ or $(1\ \ 3\ \ 2)$.

2 (a) V; $\{(1\ \ 2\ \ 3), (1\ \ 3\ \ 4), (2\ \ 4\ \ 3), (1\ \ 4\ \ 2)\}$,

$\{(1\ \ 3\ \ 2), (2\ \ 3\ \ 4), (1\ \ 2\ \ 4), (1\ \ 4\ \ 3)\}$.

(b) as in (a). Yes; $\gamma K = K\gamma$, $\forall\ \gamma \in A_4$.

3 (a) Normal.   (b) Normal.

(c) Not normal; e.g. $(1\ \ 3)\ K = \{(1\ \ 3), (1\ \ 2\ \ 3)\}$ but

$K(1\ \ 3) = \{(1\ \ 3), (1\ \ 3\ \ 2)\}$.

(d) Not normal; e.g.

$(1\ \ 3)\ T = \{(1\ \ 3), (1\ \ 2\ \ 3), (1\ \ 3\ \ 4), (1\ \ 2\ \ 3\ \ 4)\}$, but

$T\ (1\ \ 3) = \{(1\ \ 3), (1\ \ 3\ \ 2), (1\ \ 4\ \ 3), (1\ \ 4\ \ 3\ \ 2)\}$.

4 Let G be an abelian group and let H be any subgroup of G. Then for any $x \in G$,

$$xH = \{(xh: h \in H\} = \{hx: h \in H\} \quad \text{(abelian property)}$$
$$= hX$$
$$\Rightarrow H \text{ is normal in } G.$$

5 (a) $e \in C$.

(b) Suppose $c, d \in C$. Then for any $x \in G$, $cx = xc$ and $dx = xd$.

Hence $(cd)x = c(dx) = c(xd) = (cx)d = (xc)d = x(cd) \Rightarrow cd \in C$.

Also, $cx = xc \Rightarrow c^{-1}(cx)c^{-1} = c^{-1}(xc)c^{-1}$

$\Rightarrow (c^{-1}c)(xc^{-1}) = (c^{-1}x)(cc^{-1})$

$\Rightarrow e(xc^{-1}) = (c^{-1}x)e$

$\Rightarrow xc^{-1} = c^{-1}x$

$\Rightarrow c^{-1} \in C$.

(c) Let $g \in G$. Then

$$gC = \{gc: c \in C\}$$
$$= \{cg: c \in C\}, \text{ by definition of } C$$
$$= Cg.$$

Hence C is normal in G.

The centre of S_3 is $\{\iota\}$.

6 H is normal in G $\Leftrightarrow xH = Hx$, $\quad \forall\ x \in G$

$\Leftrightarrow \{x, xh\} = \{x, hx\}$, $\quad \forall\ x \in G$

$\Leftrightarrow xh = hx$, $\quad \forall\ x \in G$

$\Leftrightarrow h \in C$.

7 (a) 2; distinct left cosets are $\{2r: r \in \mathbb{Z}\}$ and $\{2r + 1: r \in \mathbb{Z}\}$.

(b) 5; distinct left cosets are $\{5r: r \in \mathbb{Z}\}$, $\{5r + 1: r \in \mathbb{Z}\}$, ...,

$\{5r + 4: r \in \mathbb{Z}\}$.

8 2, 3, 4 or 6.

(a) $\{\bar{0}, \bar{6}\}$; $\{\bar{0}, \bar{4}, \bar{8}\}$; $\{\bar{0}, \bar{3}, \bar{6}, \bar{9}\}$; $\{\bar{0}, \bar{2}, \bar{4}, \bar{6}, \bar{8}, \overline{10}\}$.

(b) $\{\iota, (1\ \ 2)(3\ \ 4)\}$; $\{\iota, (1\ \ 2\ \ 3), (1\ \ 3\ \ 2)\}$; $\{\iota, (1\ \ 2)(3\ \ 4),$ $(1\ \ 3)(2\ \ 4), (1\ \ 4)(2\ \ 3)\}$.

A_4 contains no subgroup of order 6. Hence it is not true that a group necessarily contains a subgroup of each of the orders that are possible by Lagrange's theorem.

9 Since $H \cap K \subseteq H$ and $H \cap K \subseteq K$, $H \cap K$ is also a subgroup of both H and K. Let $|H \cap K| = d$. Then by Lagrange's theorem, $d \mid |H|$ and $d \mid |K|$. Hence $d \mid \gcd(|H|, |K|)$. This gives (a) $d = 1, 2,$ or 4; (b) $d = 1$; (c) $d = 1$ or 7. In case (b), $|H \cup K| = 3 + 7 - 1 = 9$. In case (c), when $d = 1$, $|H \cup K| = 13$ and when $d = 7$, $|H \cup K| = 7$.

10 Let $g \in G$ have order m. Then by corollary 7.2.4, $m|p^2$. But G is not cyclic. Hence G contains no element of order p^2. Thus $m = p$ or $m = 1$.

11 (a) $\{\iota, (1\ \ 3), (1\ \ 2), (1\ \ 2)(1\ \ 3)\}$
$= \{\iota, (1\ \ 3), (1\ \ 2), (1\ \ 3\ \ 2)\}$.

(b) $\{\iota, (1\ \ 2), (1\ \ 2), (1\ \ 2)^2\} = \{\iota, (1\ \ 2)\}$.

(c) $\{(1\ \ 3), (2\ \ 3), (1\ \ 2\ \ 3), (1\ \ 2)(1\ \ 3), (1\ \ 2)(2\ \ 3),$
$(1\ \ 2)(1\ \ 2\ \ 3)\} = \{(1\ \ 3), (2\ \ 3), (1\ \ 2\ \ 3), (1\ \ 3\ \ 2)\}$.

Not true; K and L are both groups in (a) above, but KL is not a group.

KL contains *at most* $|K||L|$ elements. Also $|KL| \geqslant |K|$ and $|KL| \geqslant |L|$. Hence KL contains *at least* $\max\{|K|, |L|\}$ elements.

Problems 7.3a

1

x	(e, f)	(e, b)	(e, b^2)	(a, f)	(a, b)	(a, b^2)
o(x)	1	3	3	2	6	6

$G \times H$ contains an element of order $6 \Rightarrow G \times H$ is cyclic.

Since $G \times H = \langle (a, b) \rangle$ and $(\mathbb{Z}_6, \oplus) = \langle \bar{1} \rangle$, we can define an isomorphism $\phi: G \times H \to (\mathbb{Z}_6, \oplus)$ by the following table:

x	$(a, b)^0$	$(a, b)^1$	$(a, b)^2$	$(a, b)^3$	$(a, b)^4$	$(a, b)^5$
x	(e, f)	(a, b)	(e, b^2)	(a, f)	(e, b)	(a, b^2)
$\phi(x)$	$\bar{0}$	$\bar{1}$	$\bar{2}$	$\bar{3}$	$\bar{4}$	$\bar{5}$

2 Since $a^2 = e = b^2$, the Cayley table for $G \times H$ is as follows:

$*$	(e, e)	(e, b)	(a, e)	(a, b)
(e, e)	(e, e)	(e, b)	(a, e)	(a, b)
(e, b)	(e, b)	(e, e)	(a, b)	(a, e)
(a, e)	(a, e)	(a, b)	(e, e)	(e, b)
(a, b)	(a, b)	(a, e)	(e, b)	(e, e)

which clearly defines a group of type V.

3 G, H abelian

$\Rightarrow g_1 g_2 = g_2 g_1, h_1 h_2 = h_2 h_1, \forall g_1, g_2 \in G, \forall h_1, h_2 \in H$
$\Rightarrow (g_1 g_2, h_1 h_2) = (g_2 g_1, h_2 h_1),$
$\Rightarrow (g_1, h_1)(h_2, g_2) = (g_2, h_2)(g_1, h_1)$
$\Rightarrow G \times H$ is abelian.

Converse result is that if $G \times H$ is abelian, then both G and H are abelian. This is *true*. Proof by reversing the argument above.

4

x	$(\bar{0}, \bar{0})$	$(\bar{0}, \bar{1})$	$(\bar{0}, \bar{2})$	$(\bar{0}, \bar{3})$	$(\bar{1}, \bar{0})$	$(\bar{1}, \bar{1})$	$(\bar{1}, \bar{2})$	$(\bar{1}, \bar{3})$
$o(x)$	1	4	2	4	2	4	2	4

Not isomorphic: since S is abelian while D_4 is non-abelian *or* since S contains 4 elements of order 4 and 3 of order 2 while D_4 contains 2 elements of order 4 and 5 of order 2.

5 (a) 10; elements of maximum order: $(\bar{1}, \bar{1}), (\bar{1}, \bar{2}), (\bar{1}, \bar{3}), (\bar{1}, \bar{4})$.
  (b) 6; elements of maximum order: $(\bar{1}, \bar{1}), (\bar{1}, \bar{5})$.
  (c) 12; elements of maximum order: $(\bar{1}, \bar{1}), (\bar{1}, \bar{5}), (\bar{3}, \bar{1}), (\bar{3}, \bar{5})$.

6 Let $\phi \colon \mathbb{R}^* \to (U, 1) \times (\mathbb{R}^+, \times)$ be defined by
$$\phi(x) = \begin{cases} (1, x) & \text{when } x > 0; \\ (-1, -x) & \text{when } x < 0. \end{cases}$$

Clearly ϕ is a bijective mapping of $\mathbb{R}^*$ onto $U \times \mathbb{R}^+$.
Let $x, y \in \mathbb{R}^*$. Suppose first $xy > 0$.
Case (a): $x > 0$, $y > 0$. Then
$\phi(x)\phi(y) = (1, x)(1, y) = (1, xy) = \phi(xy)$.
Case (b): $x < 0$, $y < 0$. Then
$\phi(x)\phi(y) = (-1, -x)(-1, -y) = (1, xy) = \phi(xy)$.
Next suppose that $xy < 0$. We may assume w.l.o.g. that $x > 0$ and $y < 0$.
Then $\phi(x)\phi(y) = (1, x)(-1, -y) = (-1, -xy) = \phi(xy)$.
Hence in all cases $\phi(x)\phi(y) = \phi(xy)$ and ϕ is an isomorphism.

7 $\mathbb{Z}^*_{15} = \{\bar{1}, \bar{2}, \bar{4}, \bar{7}, \bar{8}\,\overline{11}, \overline{13}, \overline{14}\}$; $\langle\overline{11}\rangle = \{\bar{1}, \overline{11}\}$; $\langle\bar{2}\rangle = \{\bar{1}, \bar{2}, \bar{4}, \bar{8}\}$.
The set $S = \{a^r b^s \colon r = 0, 1; s = 0, 1, 2, 3\}$, where $a = \overline{11}$ and $b = \bar{2}$
$\qquad = (\bar{1} \otimes B) \cup (\overline{11} \otimes B)$
$\qquad = \{\bar{1}, \bar{2}, \bar{4}, \bar{8}\} \cup \{\overline{11}, \bar{7}, \overline{14}, \overline{13}\}$.
Since $S = A \times B$, the expression for each element is unique. Define
$\phi \colon \mathbb{Z}^*_{15} \to A \times B$ by $\phi(g) = (a^r, b^s)$, where $g = a^r b^s$. Then since this
expression for g is unique, ϕ is well defined and bijective. Let $g_1 = a^{r_1}b^{s_1}$ and $g_2 = a^{r_2}b^{s_2}$. Then $g_1 g_2 = a^{r_1+r_2}b^{s_1+s_2}$. Hence $\phi(g_1 g_2) = (a^{r_1+r_2}, b^{s_1+s_2})$. But $\phi(g_1)\,\phi(g_2) = (a^{r_1}, b^{s_1})(a^{r_2}, b^{s_2}) = (a^{r_1+r_2}, b^{s_1+s_2}) = \phi(g_1 g_2)$. Hence ϕ is an isomorphism. The following is a tabulation
of ϕ.

x	$\bar{1}$	$\bar{2}$	$\bar{4}$	$\bar{8}$	$\overline{11}$	$\bar{7}$	$\overline{14}$	$\overline{13}$
r, s	0, 0	0, 1	0, 2	0, 3	1, 0	1, 1	1, 2	1, 3
$\phi(x)$	$(\bar{1}, \bar{1})$	$(\bar{1}, \bar{2})$	$(\bar{1}, \bar{4})$	$(\bar{1}, \bar{8})$	$(\overline{11}, \bar{1})$	$(\overline{11}, \bar{2})$	$(\overline{11}, \bar{4})$,	$(\overline{11}, \bar{8})$

Since A is a cyclic group of order 2 and B is a cyclic group of order 4,
$A \cong (\mathbb{Z}_2, \oplus)$ and $B \cong (\mathbb{Z}_4, \oplus)$, by theorem 7.1.7.
Thus $A \times B \cong (\mathbb{Z}_2, \oplus) \times (\mathbb{Z}_4, \oplus)$ (problem 7.3b, 4).
Hence $(\mathbb{Z}^*_{15}, \otimes) \cong (\mathbb{Z}_2, \oplus) \times (\mathbb{Z}_4, \oplus)$.

Problems 7.4a

1 16; by theorem 7.4.1, $(\mathbb{Z}_{17}, \oplus)$ is cyclic and each minimal generating
set contains just one of its non-zero elements.

2 (a) Yes; (b) no, they generate the subgroup of order 3; (c) yes;
(d) no, $\{e, \alpha, \alpha^2\}$, where $o(\alpha) = 3$, is not a generating set;
9 (three sets of type (a) and six sets of type (c)).

3 (a) Each leads to a contradiction, thus: $ba = e \Rightarrow b = a^{-1} = a$;
$ba = b \Rightarrow a = e$; $ba = b^2 \Rightarrow a = b$; $ba = a \Rightarrow b = e$.
(b) When $ba = ab$ we have $(ab)^2 = a^2 b^2 = b^2$ and $(ab)^3 = (ab)b^2 = a$,
so that ab is not of order 2 or 3. Hence $o(ab) = 6$ and $G = \langle ab \rangle$.
(c) When $ba = ab^2$, $(ab)^2 = abab = a(ab^2)b = e$;
$(ab^2)^2 = ab^2 ab^2 = ab(ab^2)b^2 = a(ab^2)b^4 = e$.

4 Elements of G have order 1, 2, 4 or 8. If $G \neq C_8$, G contains no
element of order 8 and if $G \neq C_2 \times C_2 \times C_2$, not every non-identity
element of G has order 2. Hence we can find $b \in G$ of order 4.
Let $K = \langle b \rangle$ and $a \notin K$. Then $\{eK, aK\}$ is a decomposition of G
into left cosets and hence the distinct elements of G are
$\{e, b, b^2, b^3, a, ab, ab^2, ab^3\}$. Since $a^2 \in aK \Rightarrow a \in K$, contradicting

the choice of a, we have $a^2 \in eK = K$. Now $(a^2 = b)$ or $(a^2 = b^3) \Rightarrow (a^4 = b^2 \neq e) \Rightarrow o(a) = 8$, contradicting the definition of G. Hence $a^2 = e$ or b^2. Now $\{Ke, Ka\}$ is a decomposition of G into right cosets and hence $Ka = aK$. Thus $ba \in aK$. But $ba = a \Rightarrow b = e$ and $ba = ab^2 \Rightarrow a(ba)b = a(ab^2)b \Rightarrow (ab)^2 = a^2b^3$, so that $(ab)^2 = b^3$ when $a^2 = e$ and $(ab)^2 = b^5$ when $a^2 = b^2$. But in either case this implies $o(ab) = 8$, contradicting the definition of G.

(a) $G = \{a^r b^s : 0 \leqslant r \leqslant 1, 0 \leqslant s \leqslant 3\}$. Let $\phi \colon G \to \langle a \rangle \times \langle b \rangle$ be defined by $\phi(a^r b^s) = (a^r, b^s)$. Let $g_1 = a^{r_1} b^{s_1}$, $g_2 = a^{r_2} b^{s_2} \in G$. Then $ba = ab \Rightarrow g_1 g_2 = a^{r_1 + r_2} b^{s_1 + s_2}$ and hence $\phi(g_1 g_2) = \phi(g_1) \phi(g_2)$ so that ϕ is an isomorphism.

(b) $c = ab \Rightarrow c^2 = a^2 b^2$, since $ba = ab$. Hence $c^2 = b^4 = e$ and $o(c) = 2$. Then $cb = ab^2$, $cb^2 = ab^3$ and $cb^3 = ab^4 = a$. Thus $\{c^r b^s : 0 \leqslant r \leqslant 1, 0 \leqslant s \leqslant 3\}$ are the distinct elements of G, and, as in (a) above, $G \cong \langle c \rangle \times \langle b \rangle \cong C_2 \times C_4$.

(c) The Cayley table for G can be transformed into the table for the octic group given in solutions to problem 6.7a, 5 by the following substitution:

$$e = \iota, \qquad b^r = \rho_r, r = 1, 2, 3; \qquad \sigma_r = ab^{r-1}, \qquad r = 1, 2, 3, 4.$$

(d)

	I	B	B^2	B^3	A	AB	AB^2	AB^3
I	I	B	B^2	B^3	A	AB	AB^2	AB^3
B	B	B^2	B^3	I	AB^3	A	AB	AB^2
B^2	B^2	B^3	I	B	AB^2	AB^3	A	AB
B^3	B^3	I	B	B^2	AB	AB^2	AB^3	A
A	A	AB	AB^2	AB^3	B^2	B^3	I	B
AB	AB	AB^2	AB^3	A	B	B^2	B^3	I
AB^2	AB^2	AB^3	A	AB	I	B	B^2	B^3
AB^3	AB^3	A	AB	AB^2	B^3	I	B	B^2

It is clear from the table that M satisfies axioms 1, 3 and 4 for a group and since matrix multiplication is associative we may conclude that M is a group of order 8 satisfying $B^4 = I$, $A^2 = B^2 \neq I$ and $BA = AB^3$. Thus G is isomorphic to this group by identifying a with A and b with B.

Thus if G is a group of order 8, then G must be isomorphic with one of $C_2 \times C_2 \times C_2$, $C_2 \times C_4$ or C_8 if G is abelian and with D_4 or the quaternion group if G is non-abelian, giving just five types altogether.

BIBLIOGRAPHY

The following suggestions for reference and further reading are a small selection from the many algebra books available. They extend the material of this book in various directions and illustrate a number of different styles and approaches. Some are classics in the field; most are readily accessible to a first- or second-year undergraduate readership.

Algebra

Allenby, R.B.J.T. *Rings, Fields and Groups: An Introduction to Abstract Algebra*. Arnold, London, 1983.

Childs, L. *A Concrete Approach to Higher Algebra*. Springer-Verlag, New York, 1979.

Cohn, P.M. *Algebra*. Wiley, London, 1979.

Fraleigh, J.B. *A First Course in Abstract Algebra*. 3rd edn, Addison-Wesley, Reading, Mass., 1982.

Herstein, I.N. *Topics in Algebra*. 2nd edn, Wiley, New York, 1974.

Jacobsen, N. *Basic Algebra I*. 2nd edn, Freeman, San Francisco, 1985.

Lang, S. *Algebra*. Addison-Wesley, Reading, Mass., 1965.

MacLane, S. and Birkhoff, G. *Algebra*. 2nd edn, Macmillan, New York, 1979.

Sets and logic

Frege, G. *The Foundations of Arithmetic*. 2nd edn, Blackwell, Oxford, 1968.

Halmos, P.R. *Naive Set Theory*. Van Nostrand, New York, 1960. (Springer-Verlag, New York and Berlin, 1974.)

Hrbacek, K. and Jech, T. *Introduction to Set Theory*. Dekker, New York, 1978.

Stoll, R.R. *Set Theory and Logic*. Freeman, San Francisco, 1963. (Dover, New York, 1979.)

Number theory

Dudley, U. *Elementary Number Theory*. 2nd edn, Freeman, San Francisco, 1978.

Hardy, G.H., and Wright, E.M. *An Introduction to the Theory of Numbers*. 5th edn, Clarendon, Oxford, 1979.

Group theory

Hall, M., jun. *The Theory of Groups*. Macmillan, New York, 1965.

Lederman, W. *Introduction to Group Theory*. Oliver and Boyd, Edinburgh, 1973.

Macdonald, I.D. *The Theory of Groups*, Clarendon, Oxford, 1968.

Rotman, J.J. *An Introduction to the Theory of Groups*. 3rd edn, Allyn and Bacon, Boston, 1984.

Fields

Artin, E. *Galois Theory*. 2nd edn, U. of Notre Dame Press, Ind., 1959

Stewart, I. *Galois Theory*, Chapman and Hall, London, 1973.

Rings and modules

Hartley, B., and Hawkes, T.O. *Rings, Modules and Linear Algebra*, Chapman and Hall, London, 1970.

McCoy, N.H. *The Theory of Rings*. Macmillan, New York, 1964.

INDEX

trivial factorisation, 92
trivial subgroup, 187

union of sets, 10
unique factorisation, 114
unit, 93
universal quantifier, 34
universal set, 6
universal statement, 34

V, 223
Venn diagram, 10

well-ordering principle, 94

$\mathbb{Z}_n$, 145